LIVING IN SYNC

The Psychology of Technology

LIVING IN SYNC

The Psychology of Technology

Daniella Fraser

For information, permissions, or bulk orders, contact:
Raven Moon Publishing/Raven Moon Academy®
alissandra@ravenmoonacademy.com

ISBN: 978-1-7641464-1-8

First edition
Printed in Australia
2025

This book exists because of you. It is not here claiming to offer a perfect solution, but to give you something real, something that helps you feel seen, capable, and more at ease with the digital world. This book won't have all the answers, but I hope it helps you ask the right questions.

X Daniella

ABOUT THE AUTHOR

For years, Daniella has quietly empowered others to build confidence in their relationship with technology; meeting people where they are and helping them shift from overwhelm to clarity. Her ability to translate the digital world into something human and approachable has changed lives in ways that speak louder than words.

Daniella has worked in the disability sector since she was 19, quickly developing a natural aptitude for understanding complex behaviours. Over more than a decade, she's specialised in behaviour support, honing her skills through real-life experience, deep observation, and a genuine commitment to improving people's lives.

Traditional schooling was never a good fit. Undiagnosed dyslexia and ADHD made conventional learning difficult, but her neurodivergent mind offered something else: an ability to see patterns, feel nuance, and connect the dots others might miss. This intuitive way of understanding now shapes both her writing and her work.

Since 2022, she has studied transformational witchcraft, holding a Diploma and currently undertaking her advanced Diploma. Her path blends psychology, spirituality, and lived experience. Guided by empathy, creative insight, and the kind of perception that comes from living authentically, Daniella's work continues to evolve, bridging the seen and unseen in service of healing, growth, and human understanding.

https://daniellafraser.com.au

HOW THIS BOOK WILL GUIDE YOU

This book is designed to be both an exploration and a practical guide. It will help you understand the psychological and ethical implications of technology, provide strategies for cultivating mindful engagement, and offer actionable steps to create healthier digital habits. Whether you're seeking to find balance in your personal tech use, integrate ethical practices into your professional life, or design a future that values ethical technology, this book will offer you tools and insights to navigate these challenges. Throughout the chapters, you'll find a mix of theory, research, and real-world applications to empower you in making thoughtful, ethical choices about how you engage with technology.

A Note on How This Book is Laid Out

You'll notice this book is broken up into lots of smaller sections, with headings, examples and emoji's. I did this very deliberately.

As someone with dyslexia and ADHD, I know first-hand how overwhelming big blocks of text can feel. I wanted Living in Sync to be easy to dip in and out of, to revisit, and to feel more like a conversation than a lecture. It

The emojis?

They're like a marker or signposts. A way to help you feel at home on the page, just like I do when I read and write this way. It also felt right to include them, because emojis are a language we created with technology….. so, they belong here too.

CONTENTS

INTRODUCTION

If you're like the average American, you check your phone around 96 times a day. That's once every ten minutes *(Asurion, 2019)*, but just a few decades ago, that number was zero. There were no notifications vying for our attention, no endless social feeds designed to keep us scrolling, and no algorithms curating the information we consume.

Technology has evolved at an astonishing rate, transforming nearly every aspect of life; how we work, learn, connect, and even think. It has become an indispensable tool for convenience, efficiency, and global connectivity, giving us access to knowledge and opportunities like never before.

Yet, with this digital revolution comes an urgent question:

Are we using technology, or is it using us?

This book isn't here to convince you to reject technology, it's here to help you understand its profound impact and learning how to engage with it ethically, mindfully, and intentionally.

Ethical technology engagement isn't just about avoiding harm; it's about actively shaping our relationship with the digital world.

It means asking critical questions:

- How do algorithms influence our choices and beliefs?
- What responsibilities do tech companies have in safeguarding user data?
- How do we create boundaries to prevent digital burnout?

- How can we have respectful, meaningful interactions online?

If we're not conscious of these influences, we risk becoming passive consumers. Constantly reacting to the latest trends, notifications, and viral content, rather than engaging with technology in ways that align with our values. The goal of this book is to help you reclaim control over your digital life, so that technology serves you, rather than the other way around.

Throughout these pages, we will explore how technology shapes our thoughts, emotions, and decision-making, the ethical dilemmas it presents, and practical strategies for a healthier, more balanced digital experience. By the end, you'll be equipped with the tools to navigate the digital landscape with awareness and intention, ensuring that your relationship with technology supports your well-being, relationships, and greater societal good.

Technology is neither good nor bad - it's how we choose to engage with it that matters. Let 's make that choice an intentional one.

To truly engage with technology ethically, we must first understand how it affects us on a cognitive and emotional level. The relationship between the mind and digital tools is complex. Shaped by algorithms, behavioural psychology, and the way we process information online *(Fogg, 2003).*

Understanding the Psychology of Technology

To make informed choices about how we interact with technology, we must first understand its psychological effects. As *Carr (2010)* argues, technology doesn't just shape our external world—it rewires our brains, perceptions, and relationships. *(Carr, 2010)*.

Each notification ping, social media like, and algorithm-generated suggestion is carefully designed to capture and retain our attention. But at what cost?

How does this constant digital stimulation affect our ability to focus, process information, and regulate emotions?

At the core of this exploration is the recognition that our brains are naturally adaptive. Just as they adjust to physical environments, they also recalibrate in response to digital landscapes. Research suggests that the rapid pace of information consumption online can reduce attention spans, making it harder to engage deeply with complex ideas, problem-solving, or creative thought *(Rosen et al., 2013)*.

We live in an era of instant gratification, where likes, shares, and notifications provide quick dopamine hits, creating a cycle of digital dependence. While these momentary rewards can make us feel good, they also contribute to habitual scrolling, compulsive engagement, and an inability to disconnect.

The Emotional Cost of Digital Connection

Technology isn't just changing how we think, it's reshaping how we feel. Social media, in particular, plays a paradoxical role:
⚠ It aids in connection, support, and belonging, yet it can also fuel anxiety, loneliness, and self-doubt *(Twenge, 2017)*.

⚠️ It provides a platform for expression, yet it amplifies social comparison, unrealistic expectations, and self-criticism (*Fardouly et al., 2015*).
⚠️ It creates opportunities for meaningful dialogue, yet it often becomes a breeding ground for outrage and hostility *(Chou & Edge, 2012)*.

The curated nature of online spaces can distort reality, leading to unrealistic standards of success, beauty, and happiness. Constant exposure to idealised posts can leave individuals questioning their own lives, triggering negative self-perceptions and emotional exhaustion *(Chou & Edge, 2012*).

Beyond personal engagement, the collective behaviour of digital users also shapes our mental and emotional landscape. Misinformation spreads six times faster than factual news, reinforcing biases, deepening division, and fuelling emotional reactivity. Studies show that negative content is more likely to go viral, meaning that outrage often dictates what dominates our feeds, not accuracy. *(Brady et al., 2017)*.

Reclaiming Control Over Our Digital Lives

Recognising these psychological patterns gives us the power to make conscious, informed choices about how we engage with technology.

Instead of being passive consumers, we can:
✅ Set mindful boundaries for screen time and digital consumption.
✅ Curate online environments to prioritise positive, educational, and inspiring content.

✅ Develop strategies to manage digital distractions, misinformation, and emotional triggers.

When we take intentional steps toward a healthier relationship with technology, we shift from reactive engagement to purposeful interaction, ensuring that our digital tools enhance our lives rather than controlling them.

By the end of this book, you won't just understand technology's effects, you'll have the tools to actively shape your digital habits. The question is: Will you let technology dictate your behaviour, or will you take control?

CHAPTER 1:
FOUNDATIONS OF ETHICAL TECHNOLOGY AND HUMAN BEHAVIOUR

In today's rapidly evolving world, technology is more than just a tool, it has become a driving force that shapes nearly every aspect of our lives. From the way we communicate, work, and learn, to how we interact with others and perceive the world around us, technology is woven into the fabric of our daily existence. Whether it's a smartphone, social media platform, or artificial intelligence, these innovations have transformed how we experience the world.

The immediacy, reach, and convenience of technology have fundamentally altered our daily routines, and in many cases, redefined how we understand human connection and identity *(Turkle, 2015)*.

Yet, as we increasingly depend on digital systems and devices, we must pause and reflect on how we engage with these powerful tools. The rapid pace of technological advancement often leaves little time for us to consider the broader implications of our digital lives *(Carr, 2010)*.

Are we using technology in ways that serve us well, or have we become too reliant on it, compromising our mental health, privacy, and personal relationships?

Ethical technology engagement is about more than just being mindful of how we use our devices; it's about understanding the deeper consequences of technology 's omnipresence. Ethical technology engagement requires reflection on both individual digital habits and the broader societal impact of technological advancements *(Zuboff, 2019)*.

Understanding the Principles Behind the Use of Technology in Modern Life

The integration of technology into our lives isn't merely a technical or scientific issue, it's an ethical one. Every swipe, click, and connection has the potential to impact our well-being, our relationships, and society as a whole. Ethical technology use is about more than just being mindful of how we use our devices; it's about understanding the broader consequences of technology's presence in our world.

- *How do our interactions with digital tools shape our thoughts, behaviour, and emotions?*

- *How do the choices we make as individuals, communities, and societies affect the way technology is designed, deployed, and used?*

At the heart of this exploration lies the need for a framework. One that grounds our actions and decisions in principles that promote positive human flourishing. This chapter lays the groundwork for that framework, introducing key concepts and ethical considerations that will guide us throughout the book.

We will explore the foundational ideas of ethical technology use, beginning with the psychological and societal influences of technology, and moving toward the development of ethical principles that can inform our daily decisions.

The world of technology is complex, and its ethical implications are vast. My hope is that by the end of this book, you will have a clearer understanding of why it's essential to approach technology with intentionality, respect, and responsibility. These principles are not just theoretical; they have real-world applications that affect how we live, work, and relate to one another. Let's dive into the foundational ideas that will help us navigate this digital age with wisdom and integrity.

What Is Ethical Technology Engagement?
Let's face it, technology is woven into the fabric of our everyday lives. From the moment we wake up and check our phones to when we fall asleep scrolling through endless content, technology is there, shaping how we think, feel, and interact with the world.

But have you ever stopped to ask yourself:

☑ Am I engaging with technology in a way that aligns with my values?
☑ Is my technology use enhancing my life, or is it controlling me?
☑ How often do I consume digital content without questioning its impact on my mental and emotional well-being?
☑ When I share, comment, or interact online, am I contributing positively to the digital ecosystem or spreading negativity?

The Endless Buffet of Technology
Imagine you're at an all-you-can-eat buffet that never closes. The tables are packed with endless options -

some healthy, others loaded with sugar and preservatives designed to be addictive. Every time you walk by, it's hard to resist grabbing another plate, even if you 're no longer hungry. At first, the abundance is exciting. But over time, if you're not intentional about what you consume, the buffet leaves you feeling sluggish, unhealthy, and disconnected from what your body truly needs.

Technology functions much like that endless buffet. It offers incredible opportunities for connection, learning, and entertainment. Without conscious engagement, it can lead to digital overwhelm, loss of focus, and harm to our mental and emotional well-being. There's an undeniable allure in the digital space, where every scroll reveals something new, every notification tugs at our curiosity, and every algorithm suggests content tailored to our habits. However, just as eating without restraint can impact our physical health, mindless interaction with technology can harm our psychological and social health.

Example:
Henry, a 34-year-old professional who loves connecting with friends online. One evening, he opens his favourite social media app to unwind for "just five minutes." An hour later, Henry finds himself deep in a heated comment section on a news post. His mood has shifted from relaxed to angry and frustrated. Worse, when he tries to focus on reading a book afterward, his thoughts are still consumed by the digital debate. Henry's experience is not unique, it's a pattern many of us know all too well.

Ethical technology engagement also requires us to recognise that our digital actions and habits are never isolated, they form part of a broader cultural and societal ecosystem. When we engage online, we contribute to shaping digital norms and community behaviours. Each like, comment, or post plays a role in reinforcing certain values and ideas within online spaces.

This influence extends beyond social media interactions and can impact social movements, collective opinions, and even political discourse. Understanding this interconnectedness emphasises the importance of taking responsibility for our online presence and remaining conscious of the energy we bring into digital spaces. By encouraging respectful and intentional digital habits, we can contribute to a healthier online ecosystem that benefits both individuals and communities.

The question we need to ask ourselves is:
How did we get here?
The technology landscape has evolved rapidly, from simple communication tools to highly immersive digital ecosystems designed to capture and retain our attention. Social media platforms, news websites, and streaming services compete for our focus, using sophisticated algorithms to feed us content that triggers emotional responses *(Zuboff 2019)*.

This is why a casual scroll can spiral into hours of engagement, often leaving us emotionally drained rather than fulfilled.

If we're to cultivate healthier digital habits, we must start by asking the right questions:

🤔 How does our use of technology impact our well-being, relationships, and sense of self?
🤔 Are we contributing to a respectful and positive digital environment, or are we being drawn into toxic interactions?

Do we hold ourselves accountable for the way we engage online, including the content we consume and share?

Are we critical of the sources of information we encounter, or do we passively absorb and react to what we see?

These questions aren't just theoretical; they have real-world implications. Misinformation spreads rapidly online, shaping public opinion and influencing everything from elections to health decisions *(Lewandowsky, Ecker, & Cook, 2017; Vosoughi, Roy, & Aral, 2018).*

Emotional contagion, the spread of moods and emotions, also occurs in digital spaces. A single negative comment can set off a chain reaction of anger and hostility. On the flip side, positive and thoughtful engagement can inspire meaningful connections and constructive dialogue *(Kramer et al., 2014).*

Ethical Technology Engagement: A Shared Responsibility

Ethical technology engagement goes beyond setting limits for screen time. It 's about making mindful choices that promote personal well-being and contribute positively to society. It involves thinking critically about the content we consume, the messages we share, and how we navigate online spaces. Just as we are mindful of the food we put into our bodies, we must be intentional about the digital "diet" we consume daily.

However, individual action isn't enough. Tech companies play a significant role. They must design products that prioritise user privacy, mental health, and autonomy over profit-driven algorithms that exploit our attention. Governments, too, bear responsibility for creating and enforcing legislation that protects citizens from harmful digital practices.

Real change, however, happens when all of us, users, companies, and policymakers, work together. It requires a collective effort and a commitment to accountability at every level.

But how did we arrive at this point of such pervasive technological influence?

History is filled with pivotal moments when technological advancements reshaped human society. From the invention of the wheel to the development of the printing press, each innovation brought with it profound changes in how we communicated, worked, and lived. The rise of the internet and smartphones represents just the latest chapter in this ongoing evolution. Understanding this historical trajectory helps us make sense of our current digital landscape and offers valuable lessons for navigating the challenges of the future. In the next chapter, we'll delve into the history and evolution of technology's influence on society, exploring the milestones that have led us to this complex and interconnected digital age.

CHAPTER 2:
THE EVOLUTION OF TECHNOLOGY AND THE FOUNDATIONS OF ETHICAL USAGE

Technology has always been a double-edged sword. On one side, it has improved our lives in countless ways, from curing diseases to connecting people across the globe. However, on the other side, it raises tough ethical questions and challenges that we can't ignore *(Brey, 2017)*.

How do we make sure that technology serves us without causing harm?

To understand this balance, it helps to look back at key turning points in history and how they've shaped the ethical dilemmas we face today.

The Agricultural Revolution 41Settling Down and Scaling Up

Picture this: thousands of years ago, humans roamed as hunters and gatherers, living off the land. Then came the invention of farming. A radical shift that allowed people to settle in one place, grow their own food, and build communities. This revolution transformed human society by creating new challenges: how to manage resources, distribute food, and ensure that everyone gets a fair share *(Harari, 2015)*.

Though it sounds ancient, these challenges have modern echoes. Today 's tech platforms face similar dilemmas: managing content, distributing information, and even battling misinformation. Just as early farming communities struggled with resource allocation, modern digital platforms grapple with the equitable distribution of information *(Floridi, 2013).*

Stop to Ponder:
What "resources" are being controlled in the digital world today, and who decides how they 're shared?

The Printing Press: Information for the People (Sometimes)
Fast forward to the 1400s, Gutenberg's invention of the printing press transformed the way knowledge was disseminated. Suddenly, books were no longer a luxury for the elite; they became accessible to a much wider audience *(Eisenstein, 1980)*. This sparked an information revolution, much like the advent of the internet centuries later. Ideas could now be shared on a massive scale, challenging established norms and spreading knowledge far beyond what was previously imaginable.

But, as with any powerful tool, there was a catch. The very speed and reach of printed material also meant that misinformation and propaganda could spread rapidly *(Standage, 2013)*. In our digital era, social media has become the modern printing press, capable of both enlightening and misleading, depending on how it is used.

Reflection Question: How do you decide what information to trust online?

The Industrial Revolution: Machines and Moral Dilemmas

The 18th century brought about factories, mass production, and transformative inventions that reshaped the economy and society. It was a time of incredible innovation, yet it also introduced profound ethical challenges. Workers toiled in often brutal conditions, communities faced environmental degradation *(Freeman & Louca, 2001)*. The rapid industrial growth led to vast inequalities.

Today, similar challenges persist with advancements such as AI and automation. While machines now replace many factory jobs, they also encroach on sectors like finance, healthcare, and creative industries. The fundamental question remains:

Reflection Question: What happens when technological progress outpaces our ability to ensure fairness and ethical standards in society?

Real-Life Example: Think of customer service chatbots replacing human representatives. It might be efficient, but is it fair to the workers whose jobs are disappearing?

The 20th Century: From War to the Digital Dawn

The 20th century was a period of unprecedented technological acceleration, driven by global conflicts, geopolitical rivalries, and a relentless pursuit of progress. The rapid evolution of technology during this era reshaped industries, communication, and society itself, setting the stage for the digital age.

World War II: Innovation Under Pressure
During World War II, necessity drove scientific breakthroughs at an extraordinary pace. Technologies such as radar, jet engines, cryptography (Enigma and its decryption by early computers), and nuclear power not only influenced the outcome of the war but also laid the groundwork for post-war technological revolutions. The creation of Colossus, one of the world 's first programmable computers, marked the early steps toward modern computing, demonstrating how technology could process vast amounts of information at unprecedented speeds.

These wartime advancements didn't end with the war. Military research transitioned into civilian applications, influencing fields such as medicine, aviation, and information technology. For example, radar paved the way for modern air traffic control and weather forecasting, while computing breakthroughs would later evolve into personal computers and artificial intelligence.

The Space Race & The Birth of Modern Computing
The Cold War rivalry between the United States and the Soviet Union ignited another technological revolution. One that not only pushed humans beyond Earth but also accelerated digital computing and telecommunications. The launch of Sputnik in 1957 marked the dawn of the space age, forcing nations to invest in more powerful computing systems, miniaturised electronics, and satellite communication technologies *(Johnson, 2019)*.

The Apollo program, culminating in the moon landing of 1969, further drove advancements in computer processing, leading to the development of integrated circuits, a critical component in the evolution of personal computers, smartphones, and countless modern devices.

Without these early space-age innovations, the digital world as we know it would not exist.

From Industrial to Digital: The Rise of Ethical Challenges

As technological capabilities expanded, so did ethical dilemmas. The same computing systems that helped break Nazi codes in WWII would later be used for mass surveillance programs. The same rocket technologies that sent astronauts to the moon also perfected intercontinental ballistic missiles. The global spread of technology raised critical concerns about privacy, security, and the moral responsibilities of scientific discovery.

By the late 20th century, computers were no longer just tools for governments and research labs, they were becoming personal. The rise of the internet in the 1980s and 1990s connected people across the globe, but it also introduced a new era of ethical debates, from digital privacy to misinformation.

The legacy of the 20th century teaches us that technological progress is not inherently good or bad, like many things in life it depends on how it is used. The same innovations that revolutionised society also raised complex moral questions, highlighting the need for ethical frameworks that can keep pace with rapid advancements.

As we entered the digital age, the ethical concerns of the past evolved into new challenges: data privacy, artificial intelligence, digital surveillance, and the power of technology companies. The transition from 20th-century industrial advancements to the ethics of the digital world is not just a technological shift, it is a shift in responsibility, demanding that we ask: How do we ensure that innovation serves humanity rather than exploits it?

Bridging the Past and the Present: Lessons Learned

The technological revolutions of the past, from the Industrial Age to the Digital Revolution, have reshaped the world in profound ways. Yet, history teaches us that innovation without ethical consideration can lead to unintended consequences. From labour exploitation in the Industrial Revolution to modern concerns over surveillance and misinformation, each era has faced challenges that required ethical reflection and intervention.

Ethical frameworks and protective regulations have historically emerged in response to crises. Whether it was the labour reforms following the Industrial Revolution or privacy laws in response to the digital explosion. However, in today's fast-paced digital age, a reactive approach is no longer sufficient. We need to cultivate a culture of ethical foresight where developers, consumers, and policymakers work together to predict and prevent negative consequences before they become widespread problems.

By understanding history, we learn that technology should be a tool for good; a means to enhance human life without compromising our values. Ethical engagement means harnessing technology's potential while ensuring that progress does not come at the expense of social justice, privacy, or environmental sustainability *(Floridi, 2013)*.

The Digital Age and the Rise of AI: Convenience vs. Consequences

As the 20th century's breakthroughs in computing, telecommunications, and space technology set the stage for a connected world, the late 20th and 21st centuries ushered in a new kind of revolution. One defined by instant communication, vast data networks, and artificial intelligence.

The internet, smartphones, and AI-powered systems have reshaped society, allowing us to connect with anyone, access limitless information, and automate countless tasks. With these conveniences come significant challenges though, such as privacy threats, mental health concerns, and misinformation.

Privacy is under constant threat as vast amounts of data are collected, stored, and analysed. Mental health issues are on the rise as people grapple with the constant barrage of digital information and the pressures of online life.

Some studies suggest that digital distractions impact sustained focus, though research on whether attention spans are shortening remains debated. Today, we're also witnessing the rapid rise of artificial intelligence, from predictive text and personalised news feeds to complex algorithms that shape how we view the world.

Yet, the ethical use of these technologies remains a challenge. Who controls these algorithms, and how are they designed? As we embrace the benefits of the digital age, we must also confront the consequences, ensuring that innovation is balanced by responsibility.

Pause for Thought:
How often do you stop to consider the trade-offs between convenience and privacy in your daily digital interactions?

Why We Need Ethics in Tech

If there's one enduring lesson from history, it's that just because we can build something doesn't mean we should. Ethical considerations are critical in every stage of technological development, from initial design and production to everyday use. These considerations are not meant to hinder progress, but to ensure that technology enhances human well-being and upholds our shared values.

The ethical challenges we face today, from data privacy concerns to the impacts of AI on society, demand that we think through the consequences of our digital practices before they become embedded in our lives. This proactive approach is essential in a world where technology is advancing faster than our ability to regulate it.

In a sense, the evolution of technology is a mirror reflecting our societal values. The choices we make about how to engage with technology today will shape the digital landscape of tomorrow. Ethical technology is not a luxury; it is a necessity for ensuring that progress benefits everyone, rather than creating new forms of inequality or harm.

The Importance of Ethical Technology Engagement

With each historical leap forward, we've faced new ethical questions; and today 's digital landscape is no different. Here's the thing, ethics isn't just about rules, it's about values. It's about asking, "What kind of world do we want to create with technology?"

Here are some key areas where ethics come into play:

📌 Privacy: Are companies transparent about how they use your data?

📌 Mental Health: How does endless scrolling impact your well-being?

📌 Bias and Discrimination: Are algorithms fair, or do they reinforce harmful stereotypes?

📌 Accountability: Who's responsible when tech causes harm? Users, companies, or governments?

Example: Josh's Privacy Dilemma
Josh, a stay-at-home dad, has always embraced the convenience of digital technology. Like many people, he enjoys sharing updates with family and friends online, posting pictures of his kids' milestones and keeping up with loved ones who live far away. But lately, something has been bothering him. He's noticed that after searching for baby products or discussing household items near his phone, eerily specific ads appear in his social media feeds. At first, he brushed it off as coincidence. However, the more it happens, the more he starts to question- how much of his online activity is being tracked, and more importantly, where is all this data going?

Josh isn't alone in his concerns. Many people are aware that social media platforms and search engines collect data, but few fully grasp the extent of what's being tracked and how it's used. Every click, like, comment, and even the time spent hovering over a post can be logged, analysed, and monetised. The convenience of digital tools often comes at the cost of personal privacy, but the trade-offs aren't always clear until the consequences become personal. Like when a child's photo surfaces in unexpected places or when private conversations seem to shape advertising algorithms.

Josh's unease leads him to take a step back and reassess his online presence. He starts researching privacy settings, considering encrypted messaging apps, and exploring alternative platforms with stronger data protection policies.

His journey raises broader questions that many of us should consider:

How much personal information are we unknowingly sharing online?

What control do we truly have over our digital footprints?

How do we balance the benefits of connection with the need for privacy?

Josh's concerns also tie into a larger debate: as technology advances, where do we draw the line between convenience and intrusion? One of the most controversial developments in recent years is facial recognition technology.

On one hand, facial recognition has clear benefits: Helping to enhance security, prevent fraud, and even reunite missing persons with their families. However, it also raises ethical red flags. Studies have shown that these systems can be flawed, disproportionately misidentifying people of colour and leading to wrongful accusations. Additionally, widespread surveillance capabilities mean that individuals can be tracked in public spaces without their knowledge or consent.

Simple Ethical Frameworks to Think About

I know "ethical frameworks" sounds fancy, but it's really just different ways to think about what 's right and wrong. Here are a few easy ones:

1. The Greater Good (Utilitarianism):

Use it where it really helps people, like finding missing children, but limit other uses.

What decision helps the most people and causes the least harm?

Example: Should a social media platform show posts that get clicks even if they spread hate, or should it prioritise content that educates and uplifts?

2. Following the Rules (Deontology):

Only use it when people give clear and explicit consent, ensuring ethical and responsible usage.

Some things are just wrong, even if they're effective.

Example: It might be faster to build an app that doesn't ask for user consent, but it's not right.

3. Being a Good Person (Virtue Ethics):
Ask whether the software is being developed in a way that's fair and unbiased.

What would an honest, kind, and responsible person do?

Example: Think about the person who chooses to call out misinformation respectfully instead of getting into a nasty argument online.

What Can You Do?

You don't have to be a tech genius or a policymaker to engage ethically with technology.

Here's how you can start:
☑ Think before you click: Is this content healthy or helpful for me?
☑ Ask questions: What is this app or platform doing with my data?
☑ Speak up: If you see something unethical, whether it's a shady app or unfair treatment, call it out.
☑ Stay curious: Keep learning about how technology affects your life and the world around you.

Looking Ahead

As we reflect on the evolution of technology, from the dawn of agriculture and the invention of the printing press to the industrial and digital revolutions, we see a clear pattern. Each leap forward has brought tremendous opportunities but also significant ethical challenges that needed to be addressed.

The 20th century, in particular, was a turning point that laid the groundwork for the digital era. Advances driven by war, the space race, and rapid industrialisation not only expanded our technological capabilities but also reshaped societal structures and ethical norms. These advancements propelled us into an era of unprecedented digital connectivity, automation, and data-driven decision-making, forcing us to confront new dilemmas around privacy, autonomy, and accountability.

Today, as we stand on the brink of further digital innovation, ethical technology engagement isn't just about individuals making better choices, it's about companies designing tech responsibly, governments stepping in when needed, and society as a whole working together to ensure that technology serves humanity rather than exploits it. The foundations of ethical technology usage remind us that progress must be guided by a commitment to human values. Our digital future will be built on the choices we make now. Choices that balance innovation with responsibility, efficiency with fairness, and convenience with privacy.

In the next chapter, we'll explore how technology is shaping not just our societies but our very minds. We'll examine the cognitive, emotional, and behavioural impacts of our digital experiences and discuss practical strategies for navigating these challenges while encouraging a healthier, more mindful relationship with technology. The way we engage with technology today will determine whether it continues to be a tool for progress or a force of unintended consequences.

CHAPTER 3:
THE HUMAN MIND AND TECHNOLOGY

Exploring the Cognitive, Emotional, and Behavioural Effects of Technology

Technology has an uncanny ability to slip into every crevice of our lives. From the moment we wake up and instinctively check our phones to the notifications that demand our attention throughout the day, technology is not just something we use, it's a part of how we think, feel, and act. Have you ever stopped to wonder just how deeply it's shaping the way you see the world and interact with it?

It's no exaggeration to say that technology influences human cognition, emotions, and behaviour in profound ways. The very design of digital platforms is often based on psychological principles aimed at capturing and holding our attention.

Social media, for example, thrives on our need for social validation and rewards us with dopamine hits every time we get a 'like' or comment *(Alter, 2017)*. Over time, this conditioning changes our brain's response patterns, leading us to seek more of these fleeting digital highs.

How does this constant engagement with technology reshape the very wiring of our brains, and how can we take back control?

The Attention Economy and Cognitive Overload

One of the most immediate effects of technology on human thought is its impact on our attention spans. With information constantly bombarding us, our ability to focus and engage in deep, uninterrupted thought has diminished. The term "*attention economy*" refers to the way companies compete for and monetise our attention *(Williams, 2021)*. It's no wonder we often find ourselves jumping from one notification to the next, never fully present in any single task.

We used to wait in line at the grocery store, perhaps striking up a conversation with someone nearby or simply letting our thoughts wander. Now, most of us instinctively pull out our phones, scrolling through headlines or checking social media. These seemingly harmless habits add up, reinforcing a pattern of fragmented attention that impacts our ability to engage in deep thought even when we're not on our devices.

Research on neuroplasticity, the brain's ability to change in response to experiences, has revealed that technology doesn't just influence our habits; it rewires our brains. Studies suggest that constant digital multitasking may weaken the brain's capacity for sustained focus and deep thought. *(Carr, 2010 - Small, G., et al. 2020)*

When we jump from one notification to the next, we're training our neural pathways to prefer short bursts of stimulation over prolonged engagement. While this adaptability is remarkable, it also poses challenges.

The Emotional Impact of Digital Overload

We often think of technology as a tool that helps us be more efficient, but how often do we stop to consider its effect on our energy, emotions, and well-being? The reality is that the way we interact with technology shapes not just our productivity but also our mental clarity, emotional state, and ability to be present in our daily lives.

Imagine this: You wake up feeling ready for the day, but within minutes of scrolling through notifications, your mind is already scattered. One email reminds you of a task you forgot, a social media post sparks frustration, and a news headline triggers worry. Before you've even had breakfast, your energy is drained *(Newport, 2019)*.

Digital fatigue is real. Many people go through their day feeling mentally overloaded. Not because of physical exhaustion but because their attention is constantly being pulled in different directions. The expectation of instant replies, the pressure of keeping up with online trends, and the flood of constant information can leave us feeling disconnected from ourselves and our surroundings.

These days it's not just about stress; it's about how our relationship with technology mirrors our relationship with ourselves.

Think about it this way: If you were constantly demanding things from a friend, pulling them in multiple directions, never giving them time to rest, would that friendship thrive? Probably not. Yet, many of us treat ourselves (and our brains) this way daily through overconsumption of digital stimuli, lack of mental rest, and emotional burnout.

Reflection Exercise:
When was the last time you completed a task without checking your phone, email, or another digital distraction?
How did it feel? Was it easier or harder than you expected?

The truth is that technology is only as mindful as we allow it to be. While it can empower and connect us, it can also drain and distract us. Learning to navigate this balance is key to reclaiming focus and mental clarity.

The Concept of Digital Energy: How Online Engagement Affects You

Our digital interactions don't just shape our thoughts; they influence our emotional and energetic states as well. Every time we engage with online content, we are absorbing digital energy. Just like the people we surround ourselves with in real life can uplift or drain us, the media we consume has a direct impact on our mood, focus, and well-being *(Turke, 2015)*.

Some interactions bring digital nourishment, like watching an inspiring talk, engaging in meaningful discussions, or consuming content that educates and uplifts. Others, however, bring digital toxicity, such as doomscrolling negative news, engaging in online arguments, or absorbing fear-driven narratives. The more we immerse ourselves in emotionally charged digital spaces, the more they shape our internal state, often without our conscious awareness.

Have you ever noticed how reading a negative comment or seeing a distressing headline first thing in the morning can set the tone for your entire day? This isn't just a fleeting annoyance, it's a form of digital imprinting, where online experiences affect our emotional baseline, just like real-world interactions do.

The Energy Exchange Between Humans and Technology

Just as our interactions with people influence their energy and performance, the way we engage with technology affects how it functions for us. Think about how you feel when you're constantly met with negativity, stress, or frustration. Over time, this drains your energy, impacts your ability to function at your best, and may even affect your motivation or productivity. Now, apply that same concept to your relationship with technology.

Your phone, laptop, or even your printer is, at its core, an energy system. One that interacts with the energy you put into it. If every interaction you have with your devices is filled with frustration, anger, or desperation, it's not surprising that they seem to "malfunction" at the worst possible times.

A classic example of this is the printer that suddenly refuses to work right before an urgent deadline. Logically, the error might be due to a technical issue, but many of us have experienced how frustration seems to make things worse. Could it be that our energy and our state of mind affects how we engage with technology, influencing not just our perception of its performance but also the outcomes we experience?

Just as a person thrives when treated with patience and respect, our relationship with technology benefits when we shift our energy from frustration to curiosity and problem-solving. Instead of reacting with immediate irritation when something goes wrong, taking a deep breath and approaching the issue with a sense of calm and control can make all the difference.

The Frustration Loop: Technology's Ever-Changing Nature

Another key factor in this dynamic is the constant evolution of technology. Unlike skills that can be fully mastered with time, technology is always changing, which means learning and adaptation are never-ending processes. And what emotion does learning often provoke? Frustration.

The problem isn't just the technology itself, it's how we respond to it. We often expect ourselves to instantly understand new tools, updates, or systems. It's important to remember that if technology is always evolving, then frustration is an inevitable part of our experience with it. Rather than resisting this, we need to acknowledge that adaptation will always be required and that our mindset and our willingness to embrace change, determines the kind of relationship we build with technology.

Instead of striving to 'master' technology, perhaps a more realistic goal is to develop technological adaptability - the ability to remain flexible, patient, and open-minded as we navigate its everchanging landscape. When we approach technology with curiosity rather than frustration, we cultivate a healthier, more productive relationship with it.

To regain control, we must become intentional about curating our digital environment. This means identifying what drains us and what energises us and adjusting our habits accordingly.

Quick Audit Exercise:
Take a moment to reflect on your recent digital interactions.
🤔 What types of content leave you feeling inspired and motivated?
🤔 What online spaces tend to leave you feeling anxious, angry, or drained?
🤔 How can you minimise digital toxicity and cultivate more digital nourishment?

Take it a step further Exercise:
Try spending one hour without switching between tasks or apps on your phone. What did you notice about your ability to focus and stay present?

The Paradox of Hyper-fixation: When Technology Becomes an Obsession

While much attention is given to technology's role in distraction and cognitive overload, there is another, less discussed phenomenon- hyper-fixation. Unlike mindless scrolling or habitual multitasking, hyper-fixation is an intense, prolonged focus on a specific digital task, often to the exclusion of everything else. For some, this might manifest as hours spent perfecting a digital art piece, deep diving into obscure research, or meticulously organising files and data. For others, it might take the form of an all-consuming video game session or an unbroken binge-watch of tutorial videos on a newfound passion.

At first glance, hyper fixation may seem like the opposite of digital distraction as it's sustained attention, deep focus, and complete immersion. In fact, in some contexts, hyper-fixation can be beneficial, allowing individuals to achieve mastery, problem-solve in creative ways, or enter highly productive flow states. However, the paradox of hyper-fixation is that while it allows for deep engagement, it can also lead to exhaustion, neglect of other responsibilities, and an unhealthy relationship with technology.

Neuroscientifically, hyper-fixation is linked to dopamine regulation. Unlike technology addiction, which is characterised by compulsive engagement without fulfilment, hyper-fixation often feels deeply rewarding... until it isn't. The transition from deep focus to burnout can be abrupt, leaving individuals mentally drained, physically fatigued, and emotionally detached from their surroundings.

When Does Hyper-Fixation Become Harmful?

Hyper-fixation in and of itself is not inherently negative, but it becomes problematic when:

- It disrupts daily life: Hours pass without breaks for food, hydration, or basic self-care.
- It causes emotional distress: Frustration arises when external responsibilities interrupt the fixation.
- It leads to negative physical effects: Eye strain, headaches, and poor posture from prolonged screen time take a toll on well-being.
- It creates a dependency on digital immersion: Real-world interactions feel dull or unfulfilling compared to digital engagement.

This is especially prevalent in individuals with neurodivergent traits, such as those with ADHD or autism, who may experience hyper-fixation more intensely. However, in a world where digital environments are designed to maximise engagement, even neurotypical individuals can find themselves slipping into prolonged, technology-induced hyper-focus without realising it.

Breaking the Cycle: Balancing Deep Focus with Intentionality

Understanding the difference between purposeful deep work and unintentional digital hyper-fixation is key to maintaining a healthy relationship with technology. One way to navigate this is through self-imposed guardrails such as setting reminders to take breaks, using external cues (like alarms or timers), and creating structured time limits for immersion.

A simple method is the 90-minute rule, inspired by research on ultradian rhythms, which suggests that the brain works best in cycles of 90 minutes of deep focus followed by a break. Setting a timer to step away from the screen can help disrupt excessive hyper-fixation before it becomes draining.

Additionally, post hyper-fixation decompression is crucial. Transitioning out of an intense digital session can be jarring, so grounding techniques such as stepping outside, engaging in mindful breathing, or performing light physical movement can help reintegrate into real-world awareness.

Hyper-fixation, when harnessed intentionally, can be a powerful tool for creativity and productivity, but without awareness, it can become another way technology subtly takes control. Recognising the fine line between productive deep engagement and unhealthy digital fixation allows us to reclaim agency over our focus, ensuring that technology serves us, rather than the other way around.

Navigating Emotions in a Digital World

Social media can be a great tool, but it's also an emotional amplifier. Every like, comment, or curated image has the power to trigger joy, frustration, validation, or self-doubt. Seeing a friend's vacation photos might inspire happiness or spark feelings of envy. A heated comment section can quickly escalate stress levels, while a supportive online community can bring comfort and belonging. Digital interactions don't just shape our thoughts; they leave emotional imprints, influencing our moods and mental states in real time.

This unpredictability is part of what makes social media so engaging, AND so exhausting. One moment, a post might make you feel accomplished, and the next, you're questioning your worth. Studies suggest that social media triggers emotional highs and lows like those seen in gambling addiction. The rush of receiving a notification, the anticipation of whether a post will be liked, and the anxiety of waiting for a response all create a loop of emotional investment that can be mentally draining.

The Emotional Rollercoaster Effect

The brain is wired to seek validation, and the unpredictability of digital engagement fuels a cycle of emotional highs and lows:

📌 Excitement: A surge of dopamine when a post performs well.

📌 Anxiety: Uncertainty about how content will be received.

📌 Disappointment: Feeling unseen or unacknowledged if engagement is low.

📌 Comparison Stress: Measuring success or happiness against the curated posts of others.

These constant fluctuations can contribute to emotional fatigue, making it harder to regulate moods offline. Over time, the repeated exposure to digital highs and lows can lead to stress, reduced self-esteem, and even symptoms of anxiety or depression.

Stop To Ponder:
💡 How often do you check your phone for new notifications, even when you don't need to?
💡 Have you ever felt a shift in your mood after scrolling through social media?
💡 Do you find yourself feeling anxious when a message or post doesn't get a response as quickly as expected?

Recognising these emotional patterns is the first step in reclaiming control. The way we engage with technology goes beyond just checking our screen time. It's very much about emotional awareness. Understanding how digital interactions affect us on a deeper level allows us to navigate them with more intention and resilience.

Behavioural Changes and Digital Habits
Technology not only influences what we think and feel but also shapes our daily habits. From the way we communicate to how we make decisions; digital tools often guide our behaviours without us even realising it. For instance, algorithms are designed to predict and influence our preferences, subtly nudging us toward certain purchases, news articles, or content recommendations.

These behavioural patterns can become deeply ingrained, making it difficult to break free from compulsive tech habits. However, awareness is the first step toward change. By recognising the cues and triggers that lead to mindless scrolling or binge-watching, we can begin to reclaim control over our digital behaviour.

Example: Maya's Social Media Habit Hole

Maya, a 21-year-old university student pursuing a degree in environmental science. With most of her lectures and coursework online, technology is central to her education. However, the same screen she studies on also houses social media apps and streaming platforms that often lure her away from her assignments. "I'll tell myself it's just a quick scroll to clear my mind, but suddenly I 'm 30 minutes deep into a Social Media rabbit hole," Maya admits.

Her experience highlights a common struggle among students who rely heavily on digital tools for learning but face challenges maintaining focus. The accessibility of entertainment and social interaction on the same devices makes it harder to distinguish between work and play.

Practical Activity:

Identify one digital habit you'd like to change (e.g., checking your phone as you wake in the morning). Set a specific goal and track your progress over the next week. Reflect on how this change impacts your overall well-being.

Building Healthy Thought and Behaviour Patterns

The good news is that we can shape our relationship with technology in a way that supports healthy thought and behaviour patterns. It starts with setting clear boundaries and being intentional about how we engage with digital tools. This might mean scheduling screen-free time, turning off non-essential notifications, or curating your social media feed to align with your values.

Mindful tech practices can help rewire our brains for healthier habits. Simple strategies like using the Pomodoro Technique (working for 25 minutes, then taking a 5-minute break) can improve focus and productivity. Curating your social media feed to prioritise meaningful content can also reduce emotional overwhelm. Small, intentional changes add up, helping you regain control over your digital engagement.

At a broader level, it's also about advocating for ethical design practices that prioritise user wellbeing over profit. Companies that adopt these principles can help create a digital landscape that empowers rather than exploits users.

What changes can tech companies make to promote healthier digital habits?

How can individuals advocate for these changes?
As we've seen, technology is a force that influences our thoughts, behaviours, and emotional landscapes. This influence often operates subtly, shaping how we perceive the world and how we relate to others without our conscious awareness.

While personal awareness is essential, true change requires collective action. Tech companies must adopt ethical design practices that prioritise user well-being, and policymakers need to create regulations that protect mental health and privacy. As individuals, we hold the power to advocate for these changes while fostering healthier relationships with technology.

Beyond our individual efforts, there's a growing movement advocating for humane tech design. Initiatives like the Center for Humane Technology are pushing tech companies to create products that respect human attention and mental well-being. As consumers, we have the power to support these efforts by choosing platforms and tools that prioritise ethical engagement over constant monetisation of our attention.

By understanding these psychological dynamics, we can make more empowered choices that support both our personal well-being and the collective good.

CHAPTER 4:
THE COGNITIVE IMPACT OF TECHNOLOGY: ATTENTION, MEMORY, AND FOCUS

Emotional Regulation in the Digital Age

Technology has revolutionised how we think, process information, and engage with the world. In a single day, we absorb more digital content than previous generations encountered in weeks. While this unprecedented access to knowledge offers countless benefits, it also introduces cognitive challenges like shorter attention spans, an over-reliance on external memory, and constant digital distractions that weaken our ability to concentrate.

Before smartphones and constant digital access, mental endurance was a skill honed through daily habits. Memorisation was common, whether it was phone numbers, important dates, or long passages of text. People read books without skimming, solved problems without quick searches, and engaged in uninterrupted conversations for hours. Today, this level of cognitive stamina is rare, as our brains are conditioned for rapid information intake rather than deep engagement.

The shift from long-form thinking to rapid cognitive shifts affects our ability to:

- Maintain focus on demanding tasks.

- Retain and recall information.
- Think deeply and critically without distractions.

This chapter explores how technology is reshaping our cognitive abilities and what that means for our ability to focus, remember, and process complex information in an increasingly fast-paced digital world.

The Science Behind Focus Loss

Studies show that once our focus is interrupted, it takes an average of 23 minutes to regain full concentration *(Leroy, 2009)*. This phenomenon, called attention residue, means that even after switching tasks, part of our cognitive energy remains tied to the previous activity.

Attention residue is more than just an inconvenience, it directly impacts productivity and mental clarity. Even if you physically switch tasks, your brain continues processing the unfinished one in the background, reducing the mental energy available for the current task. This divided focus leads to slower decision-making, increased stress, and lower-quality output. Over time, chronic attention residue weakens our ability to engage in deep work, making tasks feel harder and more mentally draining than they should be.

Dopamine Loops & The Reward System

Every time we receive a like, comment, or notification, our brain releases dopamine, reinforcing the habit of checking our devices. Over time, this creates compulsive scrolling behaviours that make sustained focus more difficult.

Example:

From Distraction to Focus – Daniel's Study Breakthrough

Daniel, a postgraduate student, sets aside two hours for deep work. Within minutes, his phone buzzes with an email. Thinking it's urgent, he checks it, then briefly scrolls social media. One post leads to another, and before he realises it, 30 minutes have slipped away. When he finally returns to his work, his mind feels scattered, making it harder to re-engage.

Frustrated, Daniel decides to try a different approach. He activates the "Do Not Disturb" mode on his phone, places it in another room, and sets a 25-minute focus timer using the *Pomodoro Technique (Cirillo 2006)*. Instead of jumping back into his work with a cluttered mind, he spends two minutes practicing deep breathing to reset his focus.

This time, when he resumes his research, he notices a clear difference. Without the pull of notifications, he enters a state of deep concentration, completing a section of his paper without interruption. Encouraged by this progress, he commits to making mindful study habits a regular practice by limiting distractions, setting focus periods, and using brief mindfulness techniques to regain mental clarity.

Why This Matters

When we jump between digital stimuli, we remain in a state of partial attention, reducing productivity and increasing mental fatigue. Instead of fully immersing in tasks, we leave small parts of our focus behind. However, by implementing intentional focus strategies, we can train our minds to resist distractions and sustain deep work sessions.

Beyond productivity, this shift in attention patterns has emotional benefits. Instead of feeling mentally scattered and drained, individuals like Daniel regain a sense of control over their cognitive processes, reducing stress and frustration. The feeling of "always being busy but getting nothing done" becomes a massive emotional drain, contributing to anxiety and burnout. But with mindful engagement, focus can be reclaimed and strengthened.

Reflection Exercise:
💡 How often do you check your phone while working or studying?
💡 What strategies could help you regain focus after distractions?
💡 How does digital clutter impact your ability to stay engaged in deep work?

Digital Memory & The Google Effect: Are We Forgetting How to Remember?

As previously discussed, before smartphones, people relied on internal memory for remembering phone numbers, directions, or facts without external assistance. Today, our memory has been outsourced to devices, changing how we store and retrieve knowledge.

This phenomenon, called cognitive offloading, refers to delegating memory tasks to external tools (like Google or digital notes) instead of relying on recall. While this frees up mental space for complex thinking, it weakens long-term memory storage and the ability to recall information without prompts *(Risko & Gilbert 2016 - Sparrow, Liu, & Wegner 2011)*.

The Two Types of Digital Memory

🧠 Recognition Memory (Strengthened) – The ability to identify information when prompted. (E.g., Recognising a password when you see it but struggling to recall it on your own.)

🧠 Recall Memory (Weakened) – The ability to retrieve information without external cues. (E.g., Struggling to recall a friend's number from memory.)

Sadly, it goes beyond just remembering phone numbers. This over-reliance on external memory affects learning, comprehension, and problem-solving skills.

The Rise of Digital Amnesia

Studies show that people are more likely to forget information if they know they can retrieve it later. This phenomenon, known as digital amnesia, means that while we consume more content than ever before, we retain far less. The brain, always seeking efficiency, adapts by discarding details it assumes can be found with a quick search *(Sparrow, Liu, & Wegner 2011) (Kaspersky Lab 2015)*.

This has profound consequences for:

🧠 Critical thinking: If we don't deeply process information, we don't fully understand it.

🧠 Long-term memory: Information we don't actively recall fades faster.

🧠 Problem-solving skills: Creativity and innovation thrive on internal recall, not just Google searches.

Question: How much of our "knowledge" today is truly understood versus merely accessible on demand? If we lose access to our digital tools, would we still be able to think, remember, and solve problems effectively?

Cognitive Overload: The Mental Cost of Too Much Information

We live in an age of information abundance, but more knowledge doesn't always mean more understanding. Instead, digital overload creates a mental fog, making it harder to absorb, process, and retain meaningful insights.

Fact: In 2008, the average American consumed approximately 34 gigabytes of information and around 100,500 words daily. The equivalent of reading about 174 newspapers every day (Bohn & Short, 2009).

This constant influx of data results in:
🔹 Decision fatigue – The brain struggles to prioritise information.
🔹 Shorter attention spans – Skimming replaces deep thinking.
🔹 Memory fragmentation – Ideas are absorbed but not fully retained.

Example: The Endless Scrolling Effect
Imagine reading a thought-provoking article, but before reflecting on its key ideas, you immediately scroll to the next post. By the time you move on, the initial insight is already fading. This is how cognitive overload prevents deep learning.

How to Reduce Cognitive Overload:
✅ Pause before consuming more content – Let ideas settle before moving on.

✅ Use active recall – Instead of passively reading, summarise what you've learned.
✅ Schedule screen-free reflection time – Allow thoughts to fully develop before jumping into the next piece of information.

The Problem with Endless Scrolling

Our brains evolved to focus deeply on fewer things at a time, but modern technology encourages the opposite. Instead of absorbing knowledge and processing information fully, we develop a habit of constant scanning rather than deep engagement.

The Digital Consumption Trap:

Every time we scroll through short-form content, we train our brains to expect rapid stimulation.
Instead of engaging in meaningful thought, we:
📱 Absorb fragmented information without reflection.
📱 Lose the ability to sit with complex ideas.
📱 Struggle to process and retain knowledge long-term

Example:The 3-Second Attention Shift

Nick, a university student, used to enjoy reading books, but after years of consuming bite-sized content, He noticed:
📌 He couldn't focus on long articles without checking his phone.
📌 He would start an online course but get distracted by notifications.
📌 He often forgot what he had just read, needing to recheck information.

Instead of deep learning, his brain was stuck in shallow engagement mode. Constantly scanning but rarely absorbing.

How to Reverse the Scanning Habit

✅ Slow down content consumption – Pause between reading posts/articles before jumping to the next.

✅ Set reflection breaks – After watching a video or reading something interesting, take 60 seconds to summarise it in your own words.

✅ Prioritise long-form engagement – Spend more time on books, articles, or deep conversations rather than fragmented content.

Multitasking: The Illusion of Efficiency

Many people believe they are great multitaskers, juggling emails, social media, and work tasks simultaneously. But research shows that multitasking reduces productivity by up to 40% *(American Psychological Association, 2006 - Meyer & Kieras, 1997)*.

Despite popular belief, the brain does not perform multiple tasks at once. Instead, it rapidly switches focus, causing:

🔻 Slower processing speeds – It takes longer to complete tasks.

🔻 More mistakes & errors – Details are overlooked in rushed transitions.

🔻 Increased mental exhaustion – The constant switching depletes cognitive resources.

Example: The "Work-From-Home" Productivity Trap

Lisa, a marketing professional, prides herself on efficiency. She starts her day by answering emails, checking social media, and writing reports, all within the first hour. But by noon, she feels mentally drained, having accomplished far less than expected.

Why? Task-switching prevented her from reaching deep focus, making her work feel scattered and less effective.

The Fix: Lisa switched to the 90-Minute Deep Work Cycle. 90 minutes of uninterrupted focus, followed by 15 minutes of digital check-ins. Within two weeks, her productivity doubled while her stress levels dropped.

The Future of Focus & Cognitive Strength

In a world designed for distraction, reclaiming focus is an intentional choice. Every time we pause before reacting, prioritise deep work over shallow engagement, and protect our cognitive space, we rebuild our mental resilience.

Stop To Ponder:
🤔 How much of your daily digital activity is shallow engagement vs. deep, focused work?
🤔 When was the last time you read, wrote, or studied without interruptions? Do you feel mentally recharged or drained by your digital habits?

✋ Don't worry.... We won't leave you hanging without some practical strategies to reverse what's been discussed in chapter 4. Chapter 8 will provide tools to assist in our ever-evolving digital world.

CHAPTER 5:
TECHNOLOGY ADDICTION AND DOPAMINE TRAPS, MINDFUL TECHNOLOGY USE: CULTIVATING AWARENESS AND PRESENCE

Understanding Technology Addiction and the Dopamine Effect

The digital world wasn't designed for moderation, it was built for maximum engagement (Harris, *T 2019) (Newport, 2019)* *(Eyal, N. 2014)*. Every scroll, click, and like is engineered to keep us coming back for more, driven by psychological mechanics as powerful as those used in gambling and drug addiction.

Think about it: Have you ever reached for your phone just to check a message, only to find yourself still scrolling half an hour later? Or watched Netflix autoplay the next episode, even though you know you should be sleeping?

This isn't an accident. The pull you feel isn't about personal weakness or lack of willpower; it's the result of meticulously designed digital environments that exploit the brain's reward system. Understanding why technology is so addictive is the first step toward regaining control.

Dopamine is the driver. Habit loops are the mechanism. And your attention is the currency *(Lembke, 2021).*

Every time we receive a like, a notification, or unlock a new achievement in a game, our brain releases a burst of dopamine. This chemical hit reinforces the behaviour, making us want to engage with the activity again and again. It's not just the reward itself that hooks us, it's the anticipation. The uncertainty of when that next notification will come or whether our post will go viral creates a powerful psychological trap. This intermittent reinforcement is the same mechanism that fuels gambling addiction *(Skinner, B. F. 1953) (Duhigg 2012)*, making technology more addictive than we often realise.

How Technology Rewires Habit Formation

The brain operates on habit loops, a cycle of:

1. Cue (A trigger: You hear a notification ding.)
2. Craving (You feel an urge to check.)
3. Response (You open the app.)
4. Reward (You see a like, comment, or post, triggering a dopamine release.) *(Duhigg 2012)*

The more this loop repeats, the more automatic it becomes. Eventually, checking your phone isn't a choice, it's a reflex.

Tech companies intentionally design platforms to reinforce this loop, keeping you locked in a constant state of anticipation and reward-seeking. This is why notifications are unpredictable, variable rewards trigger stronger cravings, just like a slot machine.

What makes this even more insidious is that our brains quickly adapt to these dopamine bursts. Over time, the same activities provide diminishing returns, prompting us to seek more stimulation to achieve the same level of pleasure.

Have you ever found yourself scrolling through your feed, feeling less engaged than before but unable to stop? This is dopamine desensitisation at work. A phenomenon that encourages overuse while reducing genuine satisfaction.

Companies are acutely aware of these dynamics and design their platforms to exploit them. Infinite scrolling features, autoplay on streaming services, and carefully timed notifications all work together to keep users glued to their screens. Social media platforms, for example, use sophisticated algorithms to curate content that triggers emotional responses, whether it's outrage, joy, or curiosity. Each emotional spike keeps the dopamine loop active, reinforcing the cycle.

Real-World Consequences of Dopamine Dependency

The impact of this digital dopamine loop is profound. People often experience difficulty focusing on tasks that require sustained attention, as their brains become wired to seek constant stimulation. Mental health issues, including anxiety and depression, have also been linked to excessive technology use *(World Health Organization. 2019) (Twenge, J. M., & Campbell, W. K. 2018)* .

Social relationships can suffer when individuals prioritise virtual interactions over meaningful in person connections. Sleep deprivation is another common consequence, as the allure of digital content disrupts healthy bedtime routines.

Example: Elena's False Inspiration
Elena, a 29-year-old marketing consultant and passionate fitness enthusiast. She once prided herself on staying active and present during her workouts, using them to decompress from her fast paced career. Over time, her gym sessions became less about movement and more about scrolling through fitness influencers for "inspiration." What started as harmless browsing evolved into an obsessive habit of tracking likes on her workout posts and comparing herself to others online.

One afternoon, after spending nearly 30 minutes scrolling between sets, Elena realised she hadn't even broken a sweat. Worse still, her once-beloved workouts had become a dopamine-driven routine. Not for fitness, but for social validation and comparison.

What happened to Elena isn't unique. It's the same pattern driving phantom phone checking, where people instinctively reach for their devices even without notifications.

Key Takeaway: The problem isn't just "too much screen time". It's that technology rewires the way we engage with even the simplest daily tasks.

Breaking Free from the Dopamine Trap

Understanding how dopamine-driven loops work, is the first step toward taking back control. The challenge? Awareness alone isn't enough, because your brain is already wired to seek these rewards

Cognitive Rewiring & The "Now vs. Later" Struggle
The human brain isn't wired for long-term reward prioritisation. It craves immediate gratification. This is why rational awareness (I should stop scrolling) often loses against impulse (Just one more video).

This internal struggle is called time inconsistency. We overvalue immediate rewards and undervalue long-term consequences.

Example: You plan to sleep early but end up on your phone for "just 10 more minutes." Why? Because the immediate dopamine hit of entertainment overpowers the long-term consequence of feeling tired tomorrow.

Before any meaningful change happens, we need to stop blaming ourselves for lack of "willpower" and start recognising the cognitive traps at play.

Over time, constant exposure to dopamine-triggering stimuli can desensitise the brain 's reward system. This means that activities once pleasurable, such as reading a book, engaging in hobbies, or even spending time with loved ones, may lose their appeal. Additionally, technology addiction can disrupt sleep patterns, impair focus, and contribute to anxiety and depression.

Consider the story of Liam, a university student who initially downloaded a social media app to stay connected with friends. Over time, he found himself scrolling endlessly during lectures, at meals, and even while studying. Despite feeling guilty and exhausted, Liam couldn't stop. His grades began to slip, and his relationships suffered. Liam's experience highlights how technology addiction can quietly but profoundly impact various aspects of life.

Recognising the Signs of Technology Addiction

How Addicted Are You? Take This Quick Test

1. Do you feel anxious or restless without your phone?
2. Do you check your device first thing in the morning and last thing at night?
3. Have you ever lost track of time while scrolling, despite knowing you had other priorities?
4. Do you reach for your phone instinctively, even when you don't need to?
5. Have you tried to cut down screen time but found it difficult?

If you answered "yes" to 2 or more, your digital habits may be wired for dopamine reinforcement.

If you answered "yes" to 4 or more, your relationship with technology may be interfering with real world presence.

Key Takeaway: Technology addiction is about emotional reliance and cognitive conditioning, rather than just the amount of time spent online.

Dopamine Traps in Digital Design

Sneaky tech companies do more than just track engagement. They manipulate it by exploiting your dopamine responses

📌Autoplay: Keeps your brain in a passive consumption loop (ever finished an entire show without meaning to?).

📌Infinite Scrolling: Removes natural stopping points, preventing cognitive disengagement.

📌Personalised Feeds: Uses AI to anticipate cravings and show exactly what will keep you online longer.

By removing natural barriers to disengagement, platforms ensure that stopping becomes an active effort rather than a natural decision.

Mindful Technology Use: Cultivating Awareness and Presence

Mindful technology use offers a path toward healthier digital habits. Mindfulness, in this context, involves being fully present and intentional about digital interactions rather than engaging mindlessly. It's about shifting from unconscious scrolling to conscious, purpose-driven engagement.

Practical Strategies for Mindful Technology Use:

Digital Intentions: Before using technology, ask yourself, "What is my purpose for this interaction?" Set clear intentions to avoid mindless browsing.

Check-in Practices: Pause during digital interactions to assess your emotional and mental state.

If you feel drained or irritable, it may be time to disconnect.

📌 Focus on Single-Tasking: Avoid the temptation to multitask while using digital tools. Give your full attention to one task at a time.

📌 Device-Free Moments: Establish tech-free zones or times in your daily routine, such as during meals or the hour before bedtime.

📌 Mindful Scrolling: Curate your social media feeds to prioritise content that uplifts and informs rather than triggers negative emotions.
📌 Technology Gratitude Practice: Reflect on how technology positively supports your life, focusing on balance and moderation.

The Role of Self-Compassion in Digital Habits

It's essential to approach the process of mindful technology use with self-compassion. It's natural to slip into old patterns, especially in a digital environment designed to be addictive. Rather than berating oneself for excessive screen time, practice gentle self-reflection. Ask, "What was I seeking in that moment, and how can I meet that need in a healthier way?"

For instance, if scrolling through social media was an attempt to connect with others, consider calling a friend instead. If it was a way to decompress, perhaps a short walk or a creative activity might provide a similar release without the digital drain.

Cultivating Joy and Presence Beyond the Screen
One of the most rewarding aspects of mindful technology use is rediscovering joy and presence in offline activities. Engage in hobbies that don't involve screens, such as reading, gardening, or playing a musical instrument. Reconnect with nature, spend quality time with loved ones, and savour moments without the constant buzz of notifications.

Just as dopamine traps can lead to mindless engagement, mindful technology practices can rewire the brain for intentional, fulfilling experiences. Over time, the brain's reward system can recalibrate, making non-digital activities just as, if not more, rewarding.

Creating a Digital Balance That Works for You
Mindful technology use isn't about rejecting digital advancements but about creating a sustainable balance between the digital and physical realms. It's about using technology as a tool that enhances your life without overshadowing meaningful real-world experiences. By cultivating awareness and presence, we can navigate the digital world with greater intention and resilience while fully embracing the richness of offline connections and activities.

In the next chapter, we will explore practical strategies for maintaining a healthy balance between digital engagement and real-life presence, helping you thrive in both worlds.

CHAPTER 6:
LIVING BETWEEN SCREENS RECLAIMING REAL-LIFE CONNECTIONS

Today's hyper-connected world, it's all too easy to get swept up in the constant flow of information, notifications, and digital interactions. But excessive screen time can have serious consequences for both mental and physical health.

Physically, prolonged screen use can lead to what's known as "digital eye strain" or "computer vision syndrome." Symptoms include headaches, dry eyes, blurred vision, and neck pain from poor posture *(American Optometric Association (AOA)*.

Research has shown that looking at a screen for long periods without proper breaks strains the eye muscles and affects the quality of our sleep *(Chang et al. 2015)*.

The blue light emitted by screens, especially before bed, disrupts our circadian rhythms and can prevent us from getting restful sleep, leaving us feeling fatigued the next day *(Harvard Medical School 2012)*.

To combat these effects, it's crucial to take intentional breaks. Practices like the "20-20-20 rule," where every 20 minutes, you look at something 20 feet away for 20 seconds, can reduce eye strain. Additionally, ergonomic adjustments to your workspace, such as adjusting your chair height, screen level, and posture, can mitigate physical pain. Scheduling regular screen breaks, especially when working long hours, and limiting screen time before bed can also help improve overall wellbeing.

The Importance of Balance

Spending too much time in the digital world can lead to a sense of detachment from physical experiences and the people around us. Moments that should be cherished in real life, like family dinners, conversations with friends, or peaceful walks in nature, are often interrupted by endless notifications or the urge to scroll through social media. This imbalance not only strains relationships but also diminishes our ability to be present and mindful. Another example is going to see our favourite artist live in concert, then watching most (or all) of it through our phone while we record it. Even if we don't do it ourselves, we end up watching it through other people's phones or spending the night ducking and weaving, trying to see around the sea of screens. We are not able to be fully present in the moment which means lack of true fulfilment.

However, completely unplugging from the digital world isn't practical for most of us. Technology has become essential for work, education, and communication. Therefore, the goal is not to reject technology but to develop habits that encourage balanced, meaningful engagement with both digital and physical spaces.

How Digital Habits Weaken Real-World Relationships

Many people feel socially "connected" online but emotionally disconnected in real life. The ease of digital communication has weakened face-to-face skills, reducing empathy and emotional depth.

Common signs that digital habits affect real-world connections:

- Feeling distant from loved ones, even when spending time together.
- Struggling with eye contact, active listening, or deep conversations.
- Prioritising online interactions over meaningful in-person moments.

Example: Mark's Social Disconnection Cycle
Mark, a 27-year-old software developer, noticed that he and his friends often sat together scrolling their phones rather than talking. He decided to implement a simple rule: no phones at dinner. Initially, his friends resisted, but over time, they started enjoying deeper conversations, laughing more, and reconnecting beyond digital small talk.

Example: Hannah's Pitfalls of Over-Engagement in Digital Life
Hannah, a 42-year-old graphic designer, loved her digital downtime. Streaming shows, gaming, and staying active in online communities. But over time, she realised that her real-life relationships were suffering. Conversations with her partner became shorter, she declined social invitations, and despite daily digital interaction, she felt more isolated than ever.

The Shift: After noticing how tech was replacing real-world experiences, Hannah set social boundaries like, phone-free evenings, regular meetups with friends, and digital detox weekends. This shift improved her relationships and helped her reconnect with the present moment.

Strategies for Reclaiming Balance

To achieve a harmonious balance between the digital and physical realms, consider adopting the following strategies:

1. Establish Tech-Free Zones
Designate specific areas in your home where digital devices are off-limits, such as the dining table or bedroom. This encourages quality time with family and supports better sleep hygiene.

2. Schedule Screen-Free Time
Set aside daily or weekly periods when you disconnect from screens entirely. Use this time to engage in activities that nourish your mind and body, such as reading, exercising, or spending time outdoors.

3. Prioritise Face-to-Face Interactions
Make a conscious effort to engage in meaningful, face-to-face conversations with friends, family, and colleagues. Research has shown that in-person interactions boost emotional well-being and strengthen relationships.

4. Prioritise Meaningful In-Person Interactions
Rather than checking social media for validation, focus on face-to-face moments that strengthen relationships.
📌 Plan phone-free social outings (leave your devices in a box or face-down at dinner).
📌 Call instead of text. Conversations build deeper emotional bonds.

📌 When with others, practice active listening. Engage fully rather than checking notifications.

5. Set Boundaries for Digital Work
If work blurs the line between professional and personal life, establish clear working hours and stick to them. Communicate these boundaries with your employer or clients.

While technology has revolutionised the way we connect with others, it's essential not to overlook the unique value of in-person interactions. In a time when digital communication has become the norm, face-to-face interactions offer a depth and richness that can't be replicated through screens.

When we meet someone in person, we can observe their body language, hear the tone in their voice, and engage in a more meaningful, nuanced exchange. This emotional connection helps to encourage empathy and strengthen relationships.

Studies have shown that people who engage in more in-person interactions experience lower levels of stress and anxiety compared to those who rely primarily on digital communication. Physical presence encourages vulnerability, understanding, and intimacy- elements often lost in the digital space.

Moreover, offline relationships can provide the emotional support necessary to buffer against the potential negative effects of technology. While scrolling through social media might offer an illusion of connection, it often leaves us feeling more isolated and disconnected. Real-world connections, on the other hand, provide comfort and grounding, allowing us to express our emotions and engage with others in a more authentic way.

As we become more reliant on our screens for communication, it's important to make a conscious effort to unplug and reconnect with others in the physical world.

Setting aside time each day for quality in-person interactions, whether with family, friends, or even casual acquaintances, can significantly improve emotional resilience. Whether it's enjoying a meal together, going for a walk, or simply sitting down for a conversation, these moments of connection serve as a vital counterbalance to the digital overload that often dominates our lives.

The Joy of Rediscovering the Physical World

A balanced life allows us to experience the joy and wonder of the physical world. Nature walks, creative hobbies, and simple pleasures like cooking a meal from scratch become more fulfilling when we aren't distracted by the digital noise. Reclaiming these experiences can give a sense of presence, gratitude, and connection to ourselves and our surroundings.

Example: Greg's "Bob Ross" Moment
When Greg, a marketing executive, decided to dedicate his weekends to screen-free activities, he rediscovered his love for painting, a hobby he had abandoned for years. He found that immersing himself in creative expression not only reduced his stress but also reignited his inspiration at work.

Building Lasting Habits for Balance
Achieving a sustainable balance requires intention and consistency.

Here are some additional tips to help reinforce these changes over time:

✅ Track Your Screen Time: Use apps to monitor and set limits on your digital usage (Some
Devices have their own inbuilt screen time usage and apps most used)

✅ Digital Decluttering: Unsubscribe from unnecessary notifications and clean up your social media feed to reduce distractions.

✅ Family Agreements: Create agreements with family members about shared tech-free times to strengthen relationships and lead by example.

Embracing a Holistic Lifestyle

Balancing digital engagement with real-life presence calls for intentionality. When we cultivate habits that honour both realms, we unlock a more fulfilling and meaningful way of living. By approaching technology with mindfulness and integrating joyful, real-world experiences into our routines, we create a harmonious existence that nurtures both our digital and physical selves.

In the next chapter, we'll explore the profound importance of nature, physical connections, and how establishing boundaries and practicing digital minimalism can enrich our lives even further. Through these practices, we can cultivate a stronger sense of presence, peace, and harmony in an increasingly digital world.

CHAPTER 7:
RECONNECTING WITH NATURE AND THE POWER OF PHYSICAL CONNECTIONS

The Role of Nature in a Digitally Saturated World

In a world where notifications dictate our attention, and screens dominate our daily lives, it's easy to forget the power of the physical world around us. Studies have consistently shown that spending time in nature improves mental health, reduces stress, and enhances overall well-being, but as technology continues to integrate into every facet of life, people are spending less time outdoors than ever before.

Our increasing disconnection from nature has real consequences. Research indicates that excessive screen time contributes to anxiety, depression, and burnout, whereas even short exposure to natural environments lowers cortisol levels, enhances creativity, and improves focus. Despite these benefits, the modern lifestyle often places digital convenience over real-world experiences, leaving many feeling disconnected from the natural world and even their own bodies.

Recent studies reinforce these findings, demonstrating that prolonged screen exposure can alter brain structure and function, reducing grey matter in regions associated with emotional regulation and decision-making *(Stanford Center on Longevity, 2024).*

Additionally, a systematic review found that adolescents who exceeded daily screen time recommendations reported significantly higher levels of stress, fatigue, and reduced life satisfaction *(BMC Psychology, 2023).*

In contrast, exposure to natural settings has measurable positive effects. A Harvard study found that spending as little as 20 minutes in nature reduces cortisol levels, leading to a lower stress response and improved emotional resilience *(Harvard Health Publishing, 2024).*

Furthermore, a study published in PLOS ONE revealed that individuals who immersed themselves in nature for extended periods showed notable improvements in problem-solving skills, attention restoration, and overall creativity *(Atchley, Strayer, & Atchley, 2012).*

Despite the mounting evidence of nature's restorative power, modern society continues to prioritise screen-based interactions, replacing face-to-face communication and physical activity with digital alternatives. This shift has created a cycle where individuals spend less time outdoors, experience increased stress, and then seek digital distractions, further reinforcing their disconnection from natural experiences.

Breaking this pattern requires conscious integration of nature into daily life, whether through regular outdoor breaks, mindful walking, or structured screen-free time. These small but meaningful actions can restore a balance between technology use and natural engagement, ultimately fostering greater well-being and cognitive health.

One of the most effective ways to counteract digital overstimulation and reconnect with the physical world is through grounding, also known as earthing. Grounding refers to direct physical contact with the Earth's surface, such as walking barefoot on grass, soil, or sand. Emerging research suggests that this practice can have profound physiological benefits, including reduced inflammation, improved sleep quality, and lower stress levels *(Chevalier et al., 2012)*.

A study published in the Journal of Environmental and Public Health found that direct skin contact with the Earth's electrons helps stabilise the body's nervous system, promoting relaxation and restoring circadian rhythms *(Chevalier et al., 2012)*.

In addition, grounding has been linked to enhanced mood and decreased cortisol production, making it an effective counterbalance to the stress-inducing effects of excessive screen time *(Ghaly & Teplitz, 2004)*.

Integrating grounding into daily routines -whether by walking barefoot outdoors, practicing mindfulness in a natural setting, or even spending time gardening, provides a tangible way to realign with nature and counteract the mental fatigue caused by prolonged digital engagement. By incorporating grounding techniques, individuals can experience not only psychological relief but also physiological restoration, reinforcing the connection between well-being and the natural world.

Example: Sophia's Return to Real-World Joy
Sophia, a 37-year-old interior designer, lived a fast-paced, tech-heavy life. Her career required constant digital interaction. Clients sent mood boards through

Pinterest, consultations happened over Zoom, and social media was essential for business growth.

She convinced herself she didn't have time for nature. Any outdoor time she did have was often spent scrolling on her phone or replying to work emails, but despite her success, Sophia felt restless, fatigued, and creatively drained.

That changed when a friend invited her on a weekend retreat- a three-day escape to the mountains with no Wi-Fi. Reluctantly, she agreed. At first, she struggled with the lack of notifications and the urge to reach for her phone. As the weekend progressed though, she began to feel something unexpected... clarity. The simple act of walking through a forest, feeling the earth beneath her feet, and watching the sunrise brought a sense of grounding that her digital life never could.

When she returned home, she made a decision: nature had to become a priority, not an afterthought. She started scheduling morning walks, took lunch breaks outside, and even incorporated biophilic design elements (like indoor plants and natural materials) into her home and workspace.

The change was immediate. She felt mentally sharper, more creative, and deeply present in ways she hadn't in years. Sophia didn't abandon technology, but she learned that real-world experiences had to come first.

The Science Behind Nature's Healing Power

The profound impact of nature on human well-being is not 'hippie' or 'WooWoo'. It is backed by many decades of significant scientific research. Time in nature has been linked to:

✅ Lower stress levels – Green spaces significantly enhance mental well-being, reducing stress levels and promoting relaxation.
✅ Increased focus and creativity – Time outdoors enhances cognitive function and problem-solving skills.
✅ Improved physical health – Walking in nature reduces blood pressure and promotes overall cardiovascular health.
✅ Greater emotional resilience – People who regularly engage with nature report better emotional regulation and lower levels of anxiety.

Research from environmental psychology shows that time in nature activates the parasympathetic nervous system, leading to lower heart rates and increased feelings of calm *(Ulrich et al., 1991)*. Exposure to natural light also helps regulate the body's circadian rhythm, which is essential for quality sleep *(Harvard Health, 2024)*. Some studies even suggest that spending just 20 minutes outside can significantly reduce cortisol levels, proving that nature is truly a great healer and a necessity for mental well-being.

This is why practices like "*forest bathing*" (*Shinrin-yoku*), popular in Japan, have gained global recognition. The practice involves mindfully immersing oneself in a natural environment, not for exercise, but simply to exist in the moment, engaging all the senses *(Park et al., 2010)*.

The Importance of Physical Connection in a Digital Age

As technology makes communication more convenient, physical connections are increasingly being replaced by digital interactions. While virtual communication is valuable, it often lacks the depth, warmth, and emotional nuance of face-to-face interactions.

Human touch plays a vital role in emotional regulation and social bonding. Something as simple as a handshake, hug, or even sitting close to a loved one releases oxytocin, the hormone associated with trust and emotions.

This physical connection cannot be replicated through screens, no matter how advanced technology becomes. Studies have shown that people who experience regular physical contact, whether through friendship, family, or romantic relationships, tend to report higher levels of happiness and lower stress levels.

Example: Daniel's Shift from Virtual to Real-World Friendships

Daniel, a 24-year-old software engineer, thrived in digital spaces. He spent his free time gaming online with friends, chatting in Discord servers, and scrolling through Reddit. Despite being socially active in a digital sense, he felt isolated in real life.

One day, his best friend invited him to a weekly board game night at a local café. Initially, Daniel resisted, as he had an online raid scheduled, and meeting new people in person felt unnecessary when he had plenty of virtual connections. But something in him hesitated. He decided to go.

That first night was uncomfortable. Conversations felt different without an exit button, and he couldn't multitask like he did online. However, by the end of the night, he felt something he hadn't felt in years, a sense of genuine connection. Laughter wasn't just a typed "LOL," and victories weren't just on a screen. They were shared, tangible moments.

From that night on, Daniel made a conscious effort to balance his digital and physical friendships. He still enjoyed gaming, but he also committed to regular in-person meetups. Over time, his anxiety decreased, his confidence grew, and his real-world relationships deepened.

The Power of Boundaries in a Tech-Driven World

Setting boundaries with technology requires solid intention. Digital tools are incredible, but without boundaries, they can consume our time, attention, and energy.

Boundaries help create space for real-world engagement by ensuring that technology serves us, rather than the other way around.

Some effective boundary-setting strategies include:
📌 Designating tech-free zones – Keeping bedrooms and dining areas screen-free encourages real-world connection.
📌 Establishing digital curfews – Turning off screens an hour before bed promotes better sleep and relaxation.
📌 Scheduling offline time – Actively setting aside time for nature, hobbies, and in-person socialising ensures balance.
📌 Practicing "single-tasking" – Focusing on one task at a time, without digital distractions, improves mindfulness and productivity.

Example: Olivia's Weekend Reset

Olivia, a 41-year-old schoolteacher, struggled to separate work from her personal life. Emails, lesson planning, and social media updates blended into her evenings and weekends, leaving her mentally drained.

One weekend, she tried an experiment- a 48-hour digital detox. No phone, no laptop, no screens.

She planned activities she loved but often neglected, such as reading, gardening, and spending uninterrupted time with her kids.

By Sunday evening, something had shifted. Olivia felt more rested, more present, and surprisingly, more fulfilled.

After her first 48-hour detox, Olivia decided to make digital-free weekends a permanent practice. Over time, she noticed significant improvements. Her sleep quality improved, her stress levels dropped, and she became more present with her family. Colleagues even commented that she seemed more focused and less fatigued. Inspired by the transformation, she encouraged her students to try "mindful tech breaks", leading to an overall more engaged classroom environment.

Digital Minimalism: Living with Technology, Not for It

For those looking to embrace digital minimalism, small changes can make a big difference. A simple starting point is the "One-Screen Rule". This is exactly what it sounds like- only using one device at a time to avoid mindless multi-screening. Another strategy is the "Inbox Detox", where individuals set specific email-checking times instead of constantly refreshing their inbox.

Additionally, turning grayscale mode on smartphones has been shown to make social media less stimulating, reducing the urge to scroll endlessly.

Digital minimalism is intentional technology use. Meaning, focusing only on what adds value while eliminating the rest. You're not abandoning tech, you're curating a digital life that aligns with personal priorities.

Key principles of digital minimalism include:
📌 Assessing digital habits – Identifying which apps, platforms, or habits add value and which drain time and energy.
📌 Decluttering the digital space – Unsubscribing from unnecessary notifications, apps, and content.
Prioritising real-world experiences – Choosing face-to-face interactions over digital ones when possible.
📌 Inbox Detox – Regularly decluttering emails by unsubscribing from unwanted newsletters, setting up filters for important messages, and designating specific times to check emails, reducing digital overwhelm and improving focus.

By consciously choosing when and how to engage with technology, people can create a life that prioritises connection, fulfilment, and presence.

Nature, physical connection, boundaries, and minimalism are simply about reclaiming balance. The digital world is powerful, but it should never replace real-world experiences. By making conscious choices to reconnect with nature, nurture in-person relationships, set firm digital boundaries, and adopt a minimalist approach to tech, we can create a life that is rich, fulfilling, and deeply human.

With so much information competing for our attention, simplifying our digital habits isn't just helpful, it's essential. Research shows that excessive digital clutter contributes to decision fatigue, mental exhaustion, and reduced cognitive efficiency.

By reducing unnecessary digital noise, we free up cognitive space for deeper thinking, better decision-making, and more meaningful interactions. This is where digital minimalism comes in. By intentionally reducing digital clutter and prioritising meaningful engagement, we can regain focus, improve mental clarity, and create a more balanced relationship with technology.

What is one step can you take today to reconnect with the world beyond your screen?

CHAPTER 8:
DIGITAL MINDFULNESS – PERSONAL AND COLLECTIVE ACCOUNTABILITY IN DIGITAL SPACES

The Need for Digital Mindfulness

We live in an era where constant connection is the default, but is it always beneficial? Technology has rewired not just how we think (as explored in Chapter 4), but also how we feel, interact, and regulate emotions. Notifications pull us into reactive loops, doomscrolling fuels stress, and comparison drains self-worth.

What if we could change that? What if instead of mindlessly reacting to digital spaces, we engaged with intention and awareness? This is where digital mindfulness comes in.

Digital mindfulness is the practice of being aware of how digital interactions shape our emotions, attention, and relationships, then making conscious choices that align with well-being rather than harm.

In this chapter, we'll explore:

✅ The emotional impact of digital overload and emotional contagion.

✅ How to regulate online triggers and reclaim mental balance.

✅ Mindful consumption techniques to avoid exhaustion.
✅ Digital minimalism strategies to create clarity.
✅ The importance of ethical engagement and collective accountability.

The goal? To take back control. Instead of letting digital spaces dictate how we feel, we'll learn how to reshape our digital experience, one mindful choice at a time.

The Emotional Toll of Digital Overload

Digital spaces aren't neutral; they actively shape our emotions. Studies show that our moods are deeply influenced by what we consume. One negative comment can stay with us all day, while uplifting content can shift our perspective in seconds.

Unlike real-world conversations, where tone, facial expressions, and body language help us interpret meaning, online interactions remove this nuance. We are left to read between the lines, often assuming the worst.

This creates a powerful emotional ripple effect. And once negativity starts to spread, it becomes almost impossible to contain.

The Rise of Emotional Contagion

The way emotions spread online is a powerful yet often overlooked consequence of digital engagement. Emotional contagion, the phenomenon where emotions spread from one person to another, has found fertile ground in the digital age. A single negative post, an inflammatory headline, or a viral controversy can spread anger, frustration, and fear in seconds. The reverse is also true. Uplifting content has the power to inspire, motivate, and create positive change.

Unlike in-person interactions, where body language and tone provide context, digital spaces lack emotional nuance. This creates an environment where emotional responses are heightened, often leading to knee-jerk reactions, increased polarisation, and emotional exhaustion. Recognising how digital content influences emotions is key to reclaiming emotional balance.

Why Emotional Contagion is More Intense Online:

🔻 Negative emotions spread faster than positive ones.
🔻 Algorithms prioritise emotionally charged content to increase engagement.
🔻 Without tone or facial expressions, messages feel more hostile or urgent than intended.

Example: Megan's Doomscrolling Trap
Megan, a 29-year-old artist, started her mornings by checking the news. At first, it was just a habit scrolling headlines while sipping coffee. But over time, she noticed something unsettling:
📌 She felt tense, frustrated, and mentally drained before even starting her day.
📌 Her mind fixated on negative stories, even when she wasn't online.
📌 She struggled to be present, constantly refreshing for updates.
This is emotional contagion in action. The more negativity Megan consumed, the more it shaped her emotions, even when she wasn't aware of it.

What changed?
Instead of letting digital spaces dictate her mood, Megan set one simple rule: No news before breakfast. She

replaced scrolling with meditation and journaling. Within a week, she felt more calm, clear-headed, and in control. She was still informed, but now it was on her own terms.

Key Takeaway: Emotional contagion isn't just about what we read, it's about what we absorb. Choosing when and how we engage with digital spaces gives us back control over our emotional landscape.

Mindful Reflection, Ask yourself:

🤔 Does this content make me feel uplifted or drained?
🤔 Am I engaging out of genuine interest or an emotional reaction?
🤔 Is this discussion enhancing my well-being or fuelling negativity?

By being intentional with our digital engagement, we can break cycles of negativity and curate a digital experience that supports emotional well-being, instead of one that controls our emotions.

How to Break the Cycle of Emotional Contagion

Instead of absorbing digital emotions passively, we can:
✅ Pause before reacting. Ask: Does this deserve my energy?
✅ Check for credibility. Is this post factual, or designed to provoke outrage?
✅ Balance negative content with positive content. Follow uplifting, educational, and solution-oriented sources.
✅ Limit exposure to emotionally charged debates. Not every argument needs your input.

Reflection Exercise:

Think about a time when online content changed your mood negatively. How did it affect your mindset for the rest of the day? What could you have done differently?

Managing Digital Triggers & Emotional Regulation

Cyberbullying & Online Hostility: The Hidden Toll
The anonymity of online spaces has emboldened individuals to engage in cyberbullying, harassment, and toxic behaviour. Without the immediate feedback of face-to-face consequences, people often act impulsively, disregarding the real-world impact of their words.

This has led to an increase in:
🔻 Online harassment – targeted attacks, doxxing, or public shaming
🔻 Cancel culture – digital mob mentality that lacks room for discussion
🔻 Emotional distress – victims of online negativity often experience anxiety, depression, and self-doubt

Many assume cyberbullying is only a problem for teens, but adults experience it too. Online pile-on's, workplace harassment, and public shaming have left many feeling powerless in digital spaces. Unlike in-person conflicts, online hostility lingers. A single cruel comment can be read and re-read, amplifying its emotional impact.

Victims often feel trapped, unable to respond without escalating, yet unable to ignore the damage. Plus, with anonymity emboldening hostility, negative digital interactions can be just as harmful as real-world bullying.

Research indicates that cyberbullying affects adults across various age groups, leading to significant emotional distress (Australian Psychological Society, 2020).

Example: The Emotional Echo Chamber

Samantha, a 32-year-old journalist, spent hours on Twitter debating social issues. At first, she felt like she was making an impact. But over time, she noticed:

🔻 She felt drained, angry, and reactive, even in offline conversations.

🔻 She couldn't stop thinking about digital arguments, replaying them in her mind.

🔻 She was engaging with people who weren't listening, fuelling her frustration.

To regain control, Samantha implemented emotional regulation techniques:

✅ Stepping back – Muting accounts that triggered anger instead of engaging.

✅ Setting a time limit – Allowing herself 15 minutes a day for social activism instead of doomscrolling.

✅ Redirecting energy – Writing thoughtful, solution-oriented pieces instead of engaging in reactionary debates.

Over time, her emotional resilience improved, and she felt more in control of her digital experiences rather than emotionally hijacked by them.

Self-Check: Are You Caught in an Emotional Loop?

🤔 Are you more reactive than reflective in digital spaces?

🤔 Do online interactions drain your energy rather than inspire you?
🤔 Are you engaging in discussions that add value, or are they cyclical and unproductive?

Being mindful of our emotional triggers and choosing where to invest our energy allows us to engage online with clarity, rather than compulsion.

The Comparison Trap: When Social Media Fuels Insecurity
Social media is a double-edged sword. While it allows us to connect, share, and learn, it also magnifies comparison, often leaving us feeling less than. Platforms are intentionally designed to reward engagement through likes, shares, and comments, reinforcing social validation behaviours. This system can make users feel validated or invisible based on numbers rather than genuine connections.

Signs You're Stuck in the Comparison Trap:

📉 Feeling envious or inferior after scrolling social media.
📉 Constantly checking for likes, comments, or validation.
📉 Measuring self-worth against influencers, celebrities, or peers.

Example: Social Media & Self-Worth
Ethan, a fitness coach, loved following industry leaders on social media. At first, it was inspiring. But over time, he noticed a shift:

📉He felt like he wasn't doing enough, despite his success.
📉 He compared himself to influencers with curated, picture-perfect lifestyles.
📉 He started questioning if he was "falling behind."

This is the algorithm's hidden effect- it doesn't just show content, it shapes perception. Ethan felt inadequate, not because he wasn't successful, but because social media warped his perspective.

What changed?
Ethan took control by:
✅ Unfollowing accounts that fuelled comparison.
✅ Curating content that focused on real, sustainable fitness.
✅ Setting a strict limit on daily scrolling.

Within weeks, his confidence returned. He was no longer chasing an illusion; he was living his own success.

Key Takeaway: Social media isn't reality. Curating what we consume is essential for mental clarity and self-worth.

Reframing the Mindset:
Instead of "*Why don't I have what they have*?", try: "*What do I appreciate about my own life right now*?"

Instead of "*They're so much more successful than me*", try: "*What's one small step I can take toward my own goals*?"

The Reality of Curated Content

When we scroll through digital spaces, what we see often isn't the full picture. Our brains are naturally drawn to polished images of dream vacations, flawless relationships and perfect fitness journeys, without noticing the messy, complicated realities that exist behind the scenes. Research shows that screen-based visuals are processed with less sensory depth than real-world experiences, which can subtly shift how we perceive success, happiness, and even ourselves.

🌴We see Dream vacations, but not the financial stress behind them.
💪Perfectly edited fitness journeys, but not the struggles and setbacks.
❤️Flawless relationships, but not the arguments, compromises, or hard work.

Why does this matter?
Because the way our brain handles digital imagery affects both our emotions and our memories. When we process curated, surface-level snapshots without the grounding details of real life, it's easier to believe that others are effortlessly thriving while we are falling short. Neuroscience highlights that this shallow processing can distort emotional responses and reinforce unrealistic comparisons *(Smith & Johnson, 2021).*

The digital world often strips away complexity, offering polished moments without context, and in doing so, it can quietly shape how we judge ourselves and our own lives.

Reframe Your Perspective:
Instead of comparing your behind-the-scenes reality to someone's edited highlight reel, ask:
🤔 What parts of my life bring me joy?
🤔 What accomplishments am I proud of- big or small?
🤔 How can I focus on my own growth, rather than someone else's curated success?

Intentional Digital Engagement: Practicing Self-Regulation Online
Every time we click, comment, or share, we 're shaping not just our own experience but the larger digital

landscape. Yet, how often do we engage online with true awareness? Many interactions are driven by impulse rather than intention, leading to misinformation, emotional exhaustion, and unnecessary conflict.

The Mindful Engagement Rule
Before posting or responding, take a moment to ask:
🤔 Am I contributing something meaningful, or just reacting?
🤔 Would I say this in person?
🤔 Is this discussion productive, or just fuelling negativity?

Why This Matters:
Without face-to-face cues like tone and body language, misinterpretation thrives online. Taking a pause before responding can reduce conflict, encourage thoughtful discussions, and create a more positive digital space.

The Power of Algorithms: Who's Really in Control?
The content we consume isn't random; it's carefully curated to keep us engaged. Social media and news platforms leverage complex algorithms to prioritise content that maximises user engagement, often amplifying emotional responses and reinforcing cognitive biases.

If you frequently engage with outrage-driven content, your feed will become a loop of frustration. If you focus on uplifting or educational content, your digital space shifts toward inspiration.

Example: The Power of Algorithm Awareness
Jennifer, a journalist, noticed her news feed was increasingly negative. Every time she engaged with

controversial articles, her feed filled with similar content, fuelling frustration.

What she did differently:
✅ She unfollowed outrage-driven news pages.
✅ She balanced her content diet with solution-oriented journalism.
✅ She chose when to engage, rather than reacting impulsively.

The result? A calmer, more informed mindset. Jennifer realised the algorithm was shaping her emotions—but she could reshape her algorithm.

How to Take Back Control:
✅ Create a mindful media rule-limit exposure to overwhelming content.
✅ Follow sources that align with truth & well-being.
✅ Set a content review habit- unfollow accounts that no longer serve you.
✅ Balance consumption with creation. Instead of endlessly scrolling, use tech for creativity- write a blog post, create digital art, or engage in meaningful discussions.

Stop To Ponder:
Does your digital engagement energise or exhaust you? If it's draining, it's time to reassess what you interact with.

Mindful Digital Parenting – Teaching the Next Generation

The Hidden Impact of Digital Role Modelling
Children don't just listen to what parents say about technology, they watch how they use it. A parent who says, "Put your phone down," while staring at their own screen sends mixed signals. Instead, modelling intentional digital habits teaches children by example.

Practical Digital Parenting Tips:

☑ Tech-Free Zones: No devices at the dinner table or bedtime
☑ Narrate Digital Choices: "I'm putting my phone away to focus on our conversation."
☑ Balance Screen Time: Create screen-free family activities
☑ Teach Emotional Awareness: Help kids understand how online interactions impact their feelings

Example: Laura & Lily's Digital Reset
Laura noticed that both she and her 10-year-old daughter, Lily, spent evenings scrolling in silence. Instead of setting strict rules, she introduced family tech-free time, including a weekly "device-free day" where they engaged in hobbies together. Over time, Lily naturally reduced her screen time and developed healthier digital habits.

Example: Mark & Ethan's Screen-Time Struggles
Mark, a father of two, struggled to manage his 8-year-old son Ethan 's screen-time habits.
Whenever Mark asked Ethan to put down the tablet, it turned into a battle. Ethan resisted, throwing tantrums or begging for "just five more minutes." Mark soon realised that simply taking the device away wasn't solving the deeper issue, that Ethan wasn't learning how to regulate his own technology use.

Determined to shift the approach, Mark set consistent screen-time boundaries with Ethan instead of reacting in the moment. Together, they created a tech-use agreement, where Ethan could earn screen time by completing other non-digital activities first, like outdoor play, reading, or family time. Mark also made sure to participate in screen activities with Ethan, watching educational videos together instead of leaving him alone with a screen.

The results weren't immediate, but over time, Ethan became more willing to self-regulate. Knowing what to expect each day reduced tantrums, and by offering alternatives to screens, Mark helped Ethan develop more balanced habits.

Collective Digital Accountability: Cultivating Ethical Online Spaces

The Role of Individuals in Shaping Online Communities

While personal digital mindfulness is essential, it's equally important to hold digital spaces to a higher standard. Individuals and communities have the power to shape online culture by promoting:

📌 Fact-checking before sharing
📌 Calling out misinformation constructively
📌 Encouraging civil discussions over hostility
📌 Advocating for platforms that prioritise ethical tech

Example: Rebecca's Digital Activism
Rebecca, an environmental advocate, realised that social media often fuelled outrage rather than solutions. She shifted her focus from reactionary engagement to

solution-based discussions, encouraging followers to take real-world action instead of just sharing posts.

Digital Mindfulness in Professional Spaces

The expectation for employees to be reachable 24/7 has been linked to increased stress, burnout, and decreased productivity. Studies show that constant digital availability blurs the boundary between work and personal life, making it harder for individuals to fully disengage and recover. Implementing mindful tech practices can create a healthier work culture:

Workplace Digital Mindfulness Solutions:

📌 The 60-30 Rule: 60 minutes of deep focus, then a 30-minute break

📌 Notification-Free Zones: No email or Slack messages after hours

📌 Scheduled Email Times: Batch-check emails instead of constantly reacting

The Future of Mindful Digital Culture

By adopting digital mindfulness practices, we can:

✅ Improve mental well-being through conscious tech use

✅ Shift online culture from reactionary to intentional engagement

✅ Create ethical digital spaces that prioritise respect, kindness, and balance

Change starts with small, intentional steps. Whether it's setting tech boundaries, practicing emotional intelligence, or engaging in healthier online spaces, your actions shape the digital world.

What small change can you make today to reshape your digital habits?

Final Reflection: If you could make one small change to your digital habits today, what would it be?

Technology is neither good nor bad, it is a tool. How we engage with it defines our experience.

Your Digital Future Starts Now.

CHAPTER 9:
ETHICAL TECHNOLOGY INTEGRATION IN PERSONAL AND PROFESSIONAL PRACTICE

Today's world is increasingly shaped by technology every day, the line between personal life and professional duties often blurs. From the devices we use to the online platforms that connect us, every aspect of our lives is intertwined with digital tools. Yet, with the growing role of technology in our daily routines, the need for ethical engagement and mindful integration is more urgent than ever. This chapter is about applying these concepts in our personal practices and workplaces.

Whether in personal life or the workplace, ethical technology use means making intentional choices that enhance well-being, communication, and productivity. In this chapter, we move beyond theory and into practical applications: how to integrate mindful tech use into daily routines, set boundaries in digital spaces, and create ethical standards for personal and professional engagement.

Checklist for Ethical Digital Engagement
Before using a digital tool, ask yourself:
- What is my goal for using this platform?
- Does this interaction align with my values?
- Am I respecting others' privacy and consent?
- Am I engaging in a way that supports mental well-being?

Applying this framework before engaging online or adopting new digital habits creates a more thoughtful and ethical approach to technology use.

Privacy and Consent

Respecting privacy is a cornerstone of ethical technology engagement. Whether it's avoiding oversharing personal details on social media or being mindful of the data we consent to share, respecting privacy means ensuring that both our own and others' boundaries are maintained. By prioritising consent, whether it's asking before tagging someone in a post or being transparent about how our digital tools collect and use data, we create spaces that respect autonomy and promote trust.

The digital landscape is filled with privacy risks, from data breaches to AI-driven tracking. Taking control of privacy settings, using encrypted communication tools, and understanding what personal information is being collected are key steps in protecting both ourselves and others. Just as we wouldn't share a friend's personal story without permission, we should extend the same courtesy in digital spaces, whether that means not forwarding private messages or being mindful of location sharing settings.

Digital Literacy

It's essential to teach ourselves and others how to navigate the digital world responsibly. This includes understanding the platforms we use, being aware of how algorithms impact our experiences, and knowing how to protect ourselves from misinformation. Digital literacy isn't just about knowing how to use technology; it's about understanding its impact and using it in ways that align with our personal ethics.

Many online platforms use algorithms to shape what we see- news, advertisements, or social media posts. This creates an echo chamber effect, reinforcing our beliefs and limiting exposure to diverse perspectives. Developing critical thinking skills and seeking out multiple sources of information can help break out of these digital bubbles.

Fact-checking, verifying sources, and understanding how media can be manipulated (such as deepfakes or AI-generated misinformation) are now essential skills in the modern digital age. By increasing our awareness, we can make more informed decisions and engage more responsibly online.

Self-Awareness

One of the most important aspects of personal digital mindfulness is self-awareness. How does technology make us feel? Does it empower us, or does it drain our energy? Take a moment to reflect on your own habits and practices: Are you using digital tools to enhance your life, or are they contributing to stress or distraction?

Self-awareness empowers us to make conscious decisions about what to engage with, and it's the first step in ensuring that we don't lose control over our digital experiences.

A simple exercise to develop self-awareness is to track your screen time as referenced in Chapter 6. Are there apps or websites that consume more time than you realised? Do certain interactions leave you feeling frustrated or unfulfilled?

Instead of focusing on guilt, think about one small change you could make. Setting a time limit on an app, creating device-free zones, or curating a more positive online environment. By making small but intentional adjustments, we can shape our digital habits to serve us rather than control us.

Integrating Ethical Practices into the Professional World

Now, let's shift the focus to the professional realm. The digital age has transformed the workplace in ways we could not have anticipated. Remote work, virtual collaborations, digital communication platforms, and online communities have made it easier to connect, but they've also introduced new challenges in terms of professional ethics, communication, and boundaries.

Building Digital Communities with Respectful Engagement:

The professional world now extends beyond the physical office. Whether you're managing a team, collaborating with clients, or networking with peers, digital tools play an integral role in professional success. Yet, these tools must be used responsibly.

A key part of digital professional ethics is creating inclusive and respectful online environments. When working in digital teams, it's essential to set clear expectations around communication, such as respecting everyone's time by being clear about availability and response times. This helps prevent burnout and creates an environment where all voices are heard. It's about setting norms that support collaboration without overwhelming individuals.

For example, online meetings should have clear agendas to avoid unnecessary digital fatigue, and group chats should include guidelines on response expectations.

Inclusive digital spaces also require attention to accessibility, ensuring that tools and platforms accommodate diverse needs, such as captioning in video calls or alternative text for images.

In professional spaces, digital mindfulness can look like adopting practices that reduce stress and improve clarity. Consider implementing tools like structured email responses that acknowledge the need for thoughtful communication. Scheduled breaks or 'tech-free' zones at work allow teams to disconnect and recharge, enhancing productivity and mental well-being. Additionally, using platforms like Slack, Microsoft Teams, or Asana ethically, such as setting status updates to indicate availability, helps streamline communication without disrupting focus.

Respecting Boundaries in the Digital Workplace:

One of the challenges of working in a digital age is the blurred boundary between work and home life. Constant access to emails, messages, and notifications can lead to burnout. Professional digital ethics calls for setting and respecting boundaries, both for yourself and for others. This means communicating clearly about expectations for working hours, respecting colleagues' time by not contacting them outside of agreed work hours and using digital tools for focus and productivity rather than distraction.

For example, many teams have adopted asynchronous communication practices, where work and feedback are not required in real-time. This allows people to complete their tasks without the pressure of immediate responses, giving everyone the space to work when it's most productive for them and respecting personal boundaries.

Company culture plays a significant role in reinforcing these digital boundaries. Employers and team leaders can model healthy digital habits by encouraging breaks, discouraging excessive after-hours communication, and respecting personal time. By cultivating an environment where digital tools enhance work rather than overwhelm, businesses can improve both efficiency and well-being in the workplace.

Tools and Techniques for Mindful Technology Integration

To make digital mindfulness a daily practice, we need practical tools that help us stay aligned with our ethical goals. Here are a few examples that can enhance both personal and professional digital spaces:

Digital Contracts

For both personal and professional environments, creating agreements (sometimes called digital contracts) helps set expectations and boundaries. In a workplace setting, these can include guidelines for communication, availability, and the ethical use of shared digital tools.

Many companies are also implementing *right to disconnect* policies, ensuring employees are not expected to respond to work communications outside of agreed hours *(Fair Work Amendment Act, 2023; French Labour Code, 2017)*. Ethical AI guidelines and data privacy agreements can also be part of professional digital contracts, cultivating a culture of transparency and trust.

On a personal level, agreements with family or friends about limits, digital-free zones, or social media privacy settings can create balance and prevent overuse. For example, a household might establish rules about when and where phones can be used, such as keeping dinner times screen free to encourage conversation and connection.

Reducing Screen Exposure

Studies have shown that reducing screen exposure before bed improves sleep quality by regulating melatonin production, essential for a healthy sleep cycle (Christianson, 2021).

Likewise, taking breaks from technology throughout the day can lower stress levels and increase focus by reducing cognitive overload and allowing the brain to reset.

Beyond personal well-being, research indicates that limiting screen time enhances cognitive performance.

Research suggests that taking breaks from digital devices, often referred to as digital detoxes, can enhance cognitive performance, improve attention spans, and support overall mental clarity. A study by *Radtke et al. (2022)* found that digital detoxes contribute to reduced stress levels and increased well-being, particularly when structured and intentional.

Another study from the University of California, Irvine, revealed that employees who took structured breaks from digital devices experienced fewer attention lapses and improved overall productivity.

In professional spaces, encouraging employees to take breaks without digital devices, such as having a designated tech-free break room, can reduce burnout and enhance workplace well-being. The American Psychological Association (2020) highlights that constant connectivity contributes to workplace stress, whereas periodic unplugging improves problem-solving skills and emotional regulation.

Additionally, utilising focus modes or do not disturb settings on devices can help reinforce tech-free moments and prevent unnecessary distractions, creating a more balanced and intentional approach to digital engagement.

Mindful Digital Consumption

When scrolling through social media or consuming digital content, practice mindfulness by checking in with yourself: Is this content serving my mental health and personal values? Am I feeling uplifted, informed, or inspired by what I'm consuming, or is it contributing to stress, comparison, or negativity?

One way to cultivate mindful digital consumption is by setting content goals before engaging online. For example, if you log into a platform, ask yourself what you're hoping to achieve whether it's learning something new, connecting with friends, or finding inspiration. This approach prevents mindless scrolling and ensures that your digital experiences are enriching and intentional.

By integrating these tools into daily life, both individuals and organisations can encourage healthier digital habits, leading to more balanced and ethical technology engagement.

Moving Toward a More Ethical Digital Future

As technology continues to evolve, so too must our understanding of how to engage with it ethically. Digital mindfulness and ethical technology use are not static concepts; they require ongoing learning, reflection, and adaptation. By embracing these practices, we contribute to a digital world that prioritises well-being, promotes meaningful connections, and enhances communication, rather than diminishing them.

Creating an ethical digital future is a shared responsibility. As individuals, we can take small but meaningful steps, such as setting clearer boundaries, being intentional about the content we consume, and prioritising respectful digital interactions. In professional spaces, businesses and organisations can implement mindful policies that support work-life balance, privacy, and inclusivity. On a societal level, advocating for ethical technology policies and digital literacy initiatives ensures that future innovations align with human values rather than working against them.

The next time you log on, take a moment to reflect:

- Does this digital interaction align with my values?
- Am I using technology in a way that enhances my life and the lives of those around me?

By making conscious, intentional choices, we create a digital landscape that supports not just efficiency and innovation, but also humanity, respect, and well-being.

CHAPTER 10:
FROM DISTRACTION TO INTENTION: USING TECHNOLOGY FOR MEANINGFUL CHANGE

Reclaiming Control in the Digital Age
Technology has revolutionised how we work, connect, and learn, but at what cost?

Think about the last time you checked your phone. Did you open it for a specific reason, or did you get lost in an endless scroll?

The constant buzz of notifications, autoplay features, and algorithm-driven feeds have made distraction the default. Instead of technology being a tool we control, it often dictates our behaviour, pulling us away from our goals, relationships, and even our sense of self.

What if we could shift this dynamic? What if technology, instead of controlling us, became a tool for focus, self-awareness, and meaningful engagement?

This chapter is about reclaiming control.

In this chapter we will explore:
📌 Why digital distractions are so powerful (and how they hijack our brains).
📌 How to shift from passive consumption to mindful engagement.

📌 Practical ways to leverage technology for learning, connection, and personal growth.

Distraction is easy, intention takes effort; but in a world designed for consumption, living with digital intention is one of the most powerful choices we can make.
Before we can use technology more intentionally, we must first understand why it so easily hijacks our attention.

The Dopamine Loop: Why We Keep Scrolling
Every time you check your phone and see a new notification, your brain releases dopamine, a neurotransmitter linked to pleasure and reward. This creates a loop: the more you check, the more dopamine you receive, reinforcing the habit.

Tech companies are in constant competition for one thing- your attention. Social media platforms, apps, and even emails are designed to exploit psychological triggers, keeping users engaged for as long as possible.

The Cost of Distraction: When Technology Thinks for You
Distraction isn't just lost time, it's lost autonomy.

When was the last time you deliberately chose what to watch, read, or engage with online?

Most of our digital experiences are not chosen consciously, they are selected for us by AI-driven algorithms designed to predict and influence behaviour. Autoplay ensures you don't stop watching. Endless scroll keeps you engaged without an exit point. Push notifications create a sense of urgency, even when nothing is urgent.

The result? A reactive, impulse-driven digital experience, where technology dictates our focus, rather than us choosing how to engage.

If we want to reclaim control, we need to stop letting technology make decisions for us.

Example: Emma's Digital Struggle
Emma, a high school student, initially downloaded social media platforms to relax between study sessions. But over time, she noticed something strange:

Every time she opened the app for a "quick break," she lost an hour- scrolling through algorithm selected content that wasn't even relevant to her interests.

She hadn't planned to watch DIY cooking hacks or celebrity drama, but the algorithm knew what would keep her engaged.

The realisation hit her: She wasn't choosing what to watch. The platform was choosing for her.

This is the essence of algorithmic distraction, pulling us into engagement loops that override our own sense of choice.

From Passive Consumption to Active Engagement

Mindlessly consuming content is different from intentional digital interaction.

Passive Consumption: Scrolling aimlessly, watching autoplay videos, clicking on suggested content without thinking.

Active Engagement: Choosing what to watch, read, or interact with, rather than following algorithm-driven suggestions.

Research indicates that active engagement can lead to increased brand loyalty and customer satisfaction, while passive consumption may result in disengagement *(Gummerus et al., 2012).*

Example: The Digital Detox Experiment
David, an executive, felt overwhelmed by endless notifications. Instead of a temporary detox, he restructured his habits:
📌 Checking emails at set times instead of reacting immediately.
📌 Blocking distracting sites during work hours.
📌 Scheduling deep-focus sessions without digital interruptions.

The result? Digital detox strategies have been shown to significantly boost mental health, focus, and well-being.

Key Takeaway: The goal isn't to "quit" technology, it's to make every digital interaction a choice.

Leveraging Technology for Self-Improvement
When approached with intention, digital tools can enhance learning, discipline, and emotional intelligence.

Many people consume content passively, but reflecting on digital habits leads to better choices. Conscious consumption is important to practice for our own well-being.

Steps to Conscious Digital Use:
Audit screen time weekly. What apps bring value? What drains energy?
Ask: Is this activity improving my well-being or just filling time?

Create a "Mindful Digital Diet". Balance educational, creative, and entertainment content.

How Small Digital Changes Rewire the Brain
Rewiring digital habits is about restructuring how we engage.

The brain has muscle type qualities - meaning it can be trained and strengthened through use and repetition. If we train it for distraction, distraction becomes the default. If we train it for focus, intention becomes second nature.

The trick? Small, deliberate habit shifts. Before reaching for your phone in the morning, pause.

Replace doomscrolling with a 5-minute intention-setting practice. Choose what to watch before opening YouTube (instead of clicking the first suggestion).

Over time, these micro-shifts rewire the way we engage with technology.

Example: Mindful Media Consumption
Jack, a corporate professional, realised the news was stressing him out. Instead of quitting media, he:
📌 Limited news intake to 20 minutes in the morning.
📌 Followed fact-based sources instead of reactionary content.
📌 Replaced late-night scrolling with reading books on personal development.

His stress levels dropped, and he felt more informed without being overwhelmed.

There's also a variety of mindfulness-driven apps designed to track digital habits and encourage better decision-making. Some powerful options include:
📌 Rescue Time (tracks where your time goes online).
📌 Mindful Browsing (gives reminders when visiting distracting sites).
📌 Forest (gamifies focus by growing a virtual tree when you stay off your phone).

Example: Ethan's Gamer's Transformation
Ethan, a passionate gamer, struggled with balancing screen time. He wasn't willing to quit gaming but wanted to be more mindful of his habits. Using Rescue Time, he realised that what drained him wasn't gaming itself but doomscrolling forums before and after playing. By cutting unnecessary browsing and setting a gaming schedule, he found more time for exercise, work, and real-life friendships, without sacrificing his hobby.

Technology and Ethical Accountability
Beyond personal use, our digital behaviours have ethical implications. Every interaction, whether engaging in online discussions, consuming media, or supporting digital creators, shapes the online ecosystem.

The way we engage with technology impacts society, from misinformation to online discourse.

A mindful, intentional approach to technology means:

📌 Think before sharing – Is this information credible?
📌 Engage with empathy – Online interactions shape real-world perceptions

📌 Support ethical tech – Choose platforms and companies that prioritise privacy, fairness, and well-being

A Framework for Intentional Tech Use
To integrate these strategies into daily life, consider this three-step framework:

1. **Awareness: Identify Digital Triggers**
   What apps or platforms drain your time?
   What emotions arise before, during, and after use?

2. **Intention: Set Clear Digital Boundaries** Create a daily or weekly tech-use plan. Designate tech-free zones.

3. **Action: Use Technology as a Tool, Not a Distraction**
   Choose platforms that add value to your life.
   Replace passive habits (doomscrolling) with active ones (learning, creating).

The Future of Digital Intention
The digital world isn't going anywhere. But how we engage with it is a choice.

We can train our brains for distraction, or we can train them for focus. We can consume passively or engage with purpose. We can let algorithms dictate our attention or reclaim control.

Technology isn't the enemy. Lack of intention is.

Challenge: What is one intentional change you will make to your digital habits today?

How would your life change if technology worked for you, instead of the other way around?

CHAPTER 11:
CRAFTING A FUTURE OF ETHICAL TECHNOLOGY ENGAGEMENT

Designing Frameworks and Toolkits for Meaningful Digital Interaction

The landscape of technology continues to evolve at a staggering pace. From artificial intelligence to immersive virtual environments, our relationship with digital tools is more intricate than ever before. With this complexity comes responsibility. In a world where technological integration is ubiquitous, we have an opportunity, and indeed, a duty, to shape how we engage with these tools, both personally and collectively.

This chapter is not about lamenting the pitfalls of technology, but about creating intentional spaces where technology supports, nurtures, and empowers us in meaningful ways. Here, we explore how to design frameworks for ethical technology engagement, cultivate personal digital well-being, and apply ethical assessments that ensure our tech use aligns with our values. We're building the future of technology engagement. One that is grounded in ethics, intentionality, and mindfulness.

Designing Personal Digital Well-Being Toolkits

The foundation of any ethical framework is self-awareness. Before diving into larger systems and societal impacts, it's important to begin with ourselves. Personal digital well-being is about intentionally curating how we engage with technology in a way that enhances our mental, emotional, and physical well-being.

This is not a "one-size-fits-all" solution, but rather a personalised toolkit, something that evolves with your needs, goals, and values.

Step One:
Define Your Digital Priorities

The first step in creating a personal digital toolkit is to define what matters most to you in your digital life. This could include:

📌 Mental Health and Clarity: How does your technology use impact your stress levels, anxiety, or cognitive load?

📌 Social Connections: What role does technology play in fostering meaningful relationships with others?

📌 Productivity and Purpose: How does technology serve your professional and personal goals?

Physical Well-Being: Are you mindful of the time spent on screens and its impact on physical health?

By identifying your core values and priorities, you'll have a clearer sense of the boundaries and guidelines that need to be in place within your digital environment.

Step Two:
Set Clear Tech Boundaries

Once you've identified your digital priorities, it's time to set boundaries. Boundaries serve as the structure of your digital well-being toolkit. These boundaries can take many forms:

📌 Time Limits: Set limits for specific types of tech use, such as social media or screen time. Utilise apps like "Forest" or "Moment" that track and encourage healthy time usage.
📌 Tech-Free Zones: Create spaces or times where digital devices are off-limits. Dinner tables, bedrooms, and family gatherings can all be sanctuaries for undistracted human connection.
📌 Mindful Consumption: Carefully curate the content you consume. Unfollow accounts or unsubscribe from channels that contribute to negativity or unhelpful comparisons. Instead, follow accounts that inspire, educate, or bring positivity to your life.

Step Three:
Implement Daily Rituals for Digital Detox

In a world that demands constant connectivity, carving out time for a digital detox is essential. Whether it's 10 minutes in the morning or an hour before bed, these daily rituals can act as a reset for your mind, allowing you to reconnect with yourself and the world around you:

☀️ Morning Mindfulness: Start your day without your phone. Engage in a mindfulness practice, journaling, or a short meditation session before checking your devices.

🌙 Nighttime Reflection: Establish a nightly routine that's free from screens. This helps signal to your brain that it's time to wind down and sleep. Consider reading a physical book, stretching, or enjoying a hot cup of tea.

These rituals help you create a buffer between the digital world and your personal time, reducing stress and maintaining balance.

Ethical Assessments for Real-World Tech Use

Now that we've looked at how to create a digital toolkit for personal well-being, let 's explore how we can apply ethical considerations when interacting with technology on a broader scale. Ethical assessments are tools that help you determine whether your current tech use aligns with your values and the greater good.

1. Assessing Technology's Impact on Society

Start by examining the broader societal implications of the tools you 're using. Ethical engagement with technology doesn't exist in a vacuum. Every choice we make has repercussions.

Consider these questions:

? Who benefits from this technology? Does it amplify voices that are typically marginalised? Or does it reinforce inequality?

? How transparent are the companies behind this tech? Is their business model based on exploiting user data for profit?

? What impact does this tech have on the environment? How sustainable are the production and disposal practices surrounding this technology?

By considering the social, environmental, and economic impacts of the tools we use, we can make more informed, ethical choices about what technologies are worth supporting.

2. Creating an Ethical Framework for Digital Engagement

Similar to the personal toolkit, an ethical framework for digital engagement allows you to create guidelines that prioritise responsibility.

To build this framework, consider these key principles:
📌 Transparency: Ethical technology should operate with transparency. You should be fully aware of how your data is being used and the terms of engagement.
📌 Accountability: Ethical frameworks hold both users and creators accountable for their actions. This means setting guidelines for your tech use and also demanding that tech companies uphold high standards for privacy, inclusivity, and fairness.
📌 Privacy: Respecting digital privacy is a fundamental ethical consideration. Regularly audit your privacy settings and opt out of data collection where possible. Advocate for stronger privacy protections on the platforms you use.
📌 Sustainability: Ensure that the technology you use is not only environmentally friendly but that the business practices behind it are sustainable in the long term. Prioritise products, services, and platforms that adhere to these values.

Creating a framework for digital engagement means consciously aligning your everyday interactions with tech with broader social good, rather than just personal convenience.

3. Ethical Assessments for Digital Products and Platforms

Before downloading new apps, using certain platforms, or engaging with a new tech product, it's important to conduct an ethical assessment.
Here's how:
📌 Privacy and Data Usage: How does the app or platform use your data? Are they selling it? Are there clear privacy policies in place?

📌 User Experience: Does the product manipulate user behaviour in unethical ways (e.g., exploiting addiction or creating FOMO)?
📌 Social Impact: How does this tech impact society? Does it promote hate speech, divisive content, or harm in any way?

You can also design your own digital audit tool, where you rate each tech product on its ethical merit. Rate each app or platform on categories such as privacy, sustainability, and social impact to ensure you're making mindful, ethical choices.

Building a Thoughtful Digital Future
The ethical engagement with technology is not a destination; it is a journey; one that requires us to be continually mindful, intentional, and self-reflective. The frameworks and toolkits we've discussed here provide the foundation for engaging with technology in a way that enhances well-being, promotes social good, and encourages sustainable practices.

By actively applying these strategies and regularly assessing the ethical impact of our digital behaviours, we are contributing to the creation of a future where technology serves as a force for positive change rather than a source of harm.

As we design our individual toolkits and contribute to the broader conversation on ethical tech, we are creating a legacy of intentionality and accountability where digital tools are aligned with our values, support our well-being, and promote a more equitable and mindful world.

CHAPTER 12:
ETHICAL TECHNOLOGY IN ACTION – FUTURE CHALLENGES & SOLUTIONS

The Urgency of Ethical Technology
Technology is evolving at an unprecedented pace reshaping communication, business, and daily life. While innovation brings opportunity, it also raises ethical concerns.

📌 Who controls the algorithms that influence our decisions?
📌 How can we prevent AI bias from reinforcing discrimination?
📌 What safeguards ensure the metaverse doesn't become a space for exploitation?

Ethical technology isn't just about how we use devices today, it's about designing systems that will define the future.

This chapter explores:
☑️ The ethical risks of emerging technologies (AI, blockchain, metaverse)
☑️ How businesses and developers can prioritise fairness, transparency, and accountability
☑️ Practical solutions for creating ethical tech policies

By actively shaping technology rather than passively adopting it, we can ensure it remains a tool for empowerment rather than control.

1. Rethinking Transparency & Accountability in AI

AI is already making decisions about hiring, healthcare, policing, financial services, and retail/consumer interactions. But what happens when these decisions are flawed?

From personalised shopping experiences to automated customer service, AI is shaping how businesses interact with consumers, influencing what we see, what we buy, and even how we resolve complaints.

While AI-driven systems enhance convenience and efficiency, they also introduce ethical concerns, including:

📌 Bias in pricing algorithms
📌 Manipulative recommendation engines
📌 Lack of transparency in decision-making

As AI continues to expand its role in everyday life, we must ask: How do we ensure these technologies serve people fairly, rather than exploit them?

Beyond Bias: AI's Expanding Ethical Concerns

📌 AI in the Legal System – Some countries, like China, have experimented with AI judges. Should we trust machines to interpret the law?

📌 AI-Generated Misinformation – Deepfakes and AI-written articles blur the line between truth and deceptio*n*. Should AI-generated content be labelled?

📌 AI & Employment Ethics – AI automation is replacing jobs at an unprecedented rate. How do we balance innovation with job security?

📌 AI in Retail & Consumer Decisions – AI-driven recommendation engines, dynamic pricing, and automated customer service shape how we shop and interact with businesses. How do we ensure ethical transparency in AI-driven consumer influence? *(Dastin, 2018)*

Solutions for Ethical AI

📌 Bias Audits & Algorithmic Fairness – AI must be tested for discriminatory patterns before deployment.

📌 Explainable AI & Transparency – AI should provide clear reasons for its decisions, rather than operating as a "black box.".

📌 Regulating AI Accountability – If an AI system causes harm, should responsibility fall on the company, developer, or policymakers?

Global AI Ethics Standards?

AI regulation is fragmented. Europe's AI Act enforces transparency, while the U.S. and China take vastly different approaches. Should there be global AI governance to prevent ethical blind spots?

Example: AI in Hiring Bias

A major tech company implemented AI for hiring, but the algorithm unintentionally discriminated against female applicants, favouring male-coded language in résumés. Without oversight, AI can reinforce biases rather than eliminate them.

Question: How do we ensure AI is used ethically while still encouraging innovation?

As AI continues to shape the world, the ethical choices we make today will determine whether it empowers society or reinforces its greatest inequalities. The future isn't just in the hands of developers, but in those of policymakers, consumers, and ethical advocates alike.

2. Blockchain for Ethical Business & Data Ownership

Blockchain has the potential to revolutionise ethics in technology beyond cryptocurrency.

Expanding Blockchain's Ethical Applications

📌 Decentralised Identity Verification – Blockchain can help combat identity theft and fraud by providing secure, self-owned digital IDs.

📌 Combatting Misinformation – Verified blockchain records could track the origin of digital content, making it harder to spread fake news.

📌 Transparent Political Systems – Blockchain-based voting could prevent election fraud and enhance democratic participation.

Example: Blockchain Against Misinformation

Imagine a future where news articles come with a blockchain-verifiable stamp of authenticity, preventing AI-generated misinformation from spreading unchecked.

Question: Should we move towards a world where all digital content is verified through blockchain?

3. The Metaverse: Ethics of Virtual Worlds

The metaverse is more than entertainment, it is becoming an extension of real life. But ethical challenges are growing.

Beyond Privacy & Harassment: New Ethical Dilemmas

📌 Virtual Property Rights – Should digital land and assets be taxed like real estate?
📌 AI-Controlled NPCs – Should AI-driven avatars be given ethical restrictions to prevent manipulation
📌 Virtual Wealth Gaps – Will digital economies deepen existing financial inequalities in the real world?

Solutions for a Fair & Inclusive Metaverse

✓ Ethical Regulation of Virtual Economies – Gambling, digital monopolies, and exploitation must be addressed before they become systemic issues.
✓ Transparent Data Policies – Users should have full control over their VR-tracked data.
✓ AI Moderation – AI should be used to detect and prevent metaverse-based harassment in real time *(Williams et al., 2021)*.

In 2022, Meta's VR platform faced backlash when users reported harassment within virtual spaces.

This raised urgent questions: Who is responsible for moderating VR environments? Should AI step in when human moderators can't?

Question: If virtual assets become as valuable as real-world ones, should they be legally protected in the same way?

4. Future-Proofing Ethical Tech Policies

To prevent ethical issues before they arise, we must proactively shape policies for emerging technologies.

What Governments Can Do:

📌 AI Fairness Laws – Enforce transparency in AI-driven decisions.

📌 Decentralised Data Rights – Recognise personal data as user-owned property, rather than corporate assets.
📌 Metaverse Consumer Protection – Prevent virtual world exploitation before the industry expands.

What Companies Can Do:
📌 Ethical AI & Tech Oversight Boards – Internal committees dedicated to responsible innovation.
📌 Eco-Friendly AI Models – Prioritise sustainable energy use in AI training.
📌 Blockchain-Based Accountability – Ensure transparency in digital content, business practices, and data transactions.

What Individuals Can Do:
📌 Support Ethical Tech Companies – Prioritise products that respect privacy and transparency. Demand Accountability – Advocate for policies that ensure fair and responsible tech.
📌 Stay Informed – Keep up with AI and digital ethics trends to make conscious decisions about your digital footprint.

Final Thought: As with everything, prevention is better than a cure. So, ethical technology is not just about fixing problems, we need to be preventing them before they arise. Are we ready to take that responsibility?

Taking Responsibility for Ethical Technology

? Will we allow AI to make unchecked decisions?

? Can we create fair digital economies where users control their own data?

? How do we ensure that virtual worlds are safe and inclusive?

? The answers to these questions will define the digital age. And it's up to us to demand ethical, transparent, and responsible technology.

The future of ethical technology starts with awareness, action, and accountability. Are we ready to shape it?

CONCLUSION

The Future of Technology Is in Our Hands

As stated, numerous times throughout this book, technology is neither good nor bad, it is a tool. Melvin Kranzberg famously stated, "Technology is neither good nor bad; nor is it neutral"
(Kranzberg, 1986). Just like a gun can't be fired with influence. A tool that reflects the intent of its users, the ethics of its creators, and the habits of its consumers. Throughout this book, we have explored how digital engagement shapes everything from our cognitive abilities and emotional resilience to our relationships and societal well-being.

In a world where technology is designed to capture attention, influence emotions, and drive behaviour, the real question is: How do we take control?

The answer lies in intentionality.

Every time we scroll, click, engage, or disconnect, we make a choice. We can either let technology dictate our habits, or we can engage with awareness. We can either consume passively, or we can create mindfully. We can either fuel division and toxicity, or we can shape digital spaces that encourage understanding, accountability, and connection.

Your Role in Shaping a Mindful Digital Future

The challenge ahead is not to abandon technology, it is to learn how to live in sync with it.

This means:

✅ Choosing deep focus over constant distraction.

✅ Prioritising human connection over digital validation.
✅ Curating a digital environment that nourishes rather than drains you.
✅ Engaging in ethical, solution-focused conversations instead of reactionary outrage.
✅ Holding tech companies accountable for transparency, accessibility, and ethical innovation.

Each of us has a role to play in shaping the future of digital culture. As *Henry Jenkins (2006)* explains, digital spaces are shaped by participatory culture. Meaning that what we tolerate becomes the standard, what we engage with becomes the algorithm, and what we create becomes the next wave of content.

If enough of us shift toward mindful digital habits, we set the foundation for a healthier, more ethical technological landscape. Not just for ourselves, but for future generations.

A Call to Action: What's Your Next Step?

This book was never meant to be just information or a passive intellectual exercise. It's a guide for taking action.

So now, it's your turn:

✅ What's one small change you can make today to realign your digital habits with your values?

✅ What online behaviours will you challenge, within yourself and within your communities? How will you use technology as a tool for growth rather than distraction?

Starting today, pause before your next digital interaction. The next notification you receive... will you react out of habit or choose to engage with intention? The next online debate... will you contribute to the noise or foster meaningful dialogue? The next algorithmic recommendation... will you consume mindlessly or critically evaluate its impact?

Remember, change doesn't happen overnight, but it does happen with conscious choices, repeated daily.

Technology isn't going anywhere, but the way we engage with it is entirely up to us.

So the question remains:

Will you shape your digital habits, or will you let them shape you?

Appendix 1:

AT HOME QUIZ

1. How often do you check your phone Without a specific purpose?

a) Rarely

b) Sometimes

c) Frequently

d) Constantly

Actionable Tool: Set a designated "check-in" time for your phone instead of mindlessly reaching for it. Use apps that track screen time to bring awareness to your habits.

2. ***In what ways does technology enhance your daily life? (Select all that apply)***

a) Helps me stay connected with loved ones

b) Provides quick access to information

c) Assists with productivity and organisation

d) Offers entertainment and relaxation

Actionable Tool: Identify which benefits align with your values and create intentional routines to maximise them. For example, schedule specific time slots for productivity and relaxation rather than letting tech dictate your day.

Section 2: Identifying Digital Stressors

3. ***Which of the following digital habits makes you feel overwhelmed or drained?***

a) Excessive notifications and constant checking of messages

b) Comparing myself to others on social media

c) Feeling pressured to always be available online

d) Consuming too much negative news or content

Actionable Tool: Customise your notifications and unfollow accounts that don't bring value to your life. Try a "digital detox day" to reset your online habits.

4. When do you feel most in control of your technology use?

a) When I set time limits and boundaries

b) When I take regular digital detoxes

c) When I use tech to support my goals rather than distract me

d) When I consciously curate my digital environment

Actionable Tool: Use "Do Not Disturb" modes, schedule downtime, and create a tech-free morning or evening routine to reinforce balance.

Section 3: Mindful Tech Engagement

5. How do you currently manage digital distractions in your life?

a) I turn off non-essential notifications

b) I schedule screen-free time daily

c) I track my screen time and adjust my habits accordingly

d) I struggle with managing distractions

Actionable Tool: Implement the Pomodoro technique: Work for 25 minutes, then take a 5minute break without screens. Consider placing your phone in another room while focusing.

6. *What is one change you could make to improve your digital well-being?*

a) Create tech-free zones in my home

b) Be more intentional with social media use

c) Replace screen time with offline activities
d) Practice digital mindfulness, such as pausing before reacting online

Actionable Tool: Start a "screen-free" challenge—commit to at least one meal or evening a day without screens and track how it affects your well-being.

Section 4: Creating a Balanced Digital Lifestyle

7. ***How do you prioritise real-life connections over digital interactions?***
a) I schedule time for in-person activities
b) I put my phone away during conversations
c) I engage in hobbies that don't involve screens
d) I could improve in this area

Actionable Tool: Try the "Phone Stack" rule when with friends or family—everyone places their phone in a stack, and the first to check it picks up the tab or performs a small forfeit.

8. ***What steps could you take to create a healthier relationship with technology?***
a) Limit my digital consumption to align with my values
b) Take regular breaks from screens
c) Be mindful of how digital content affects my emotions
d) Set clear boundaries between online and offline life

Actionable Tool: Use a habit tracker to note screen time reductions and reflect on how you feel after reducing certain tech interactions.

Bonus Reflection Question: What is one small, immediate step you can take today to create a more intentional and tech-aware lifestyle?

Final Actionable Tool: Write down your answer and set a reminder to check in with yourself in a week. See how the change has impacted your daily life and adjust accordingly.

GLOSSARY

20-20-20 Rule – A guideline for reducing eye strain that suggests taking a break every 20 minutes to look at something 20 feet away for 20 seconds.

A

Active Engagement – The intentional and mindful interaction with digital content, where the user makes conscious choices about what to watch, read, or engage with.

AI-Driven Algorithms – Artificial intelligence systems that analyse user behaviour and predict content that will maximise engagement and retention.

AI-Driven Tracking – The use of artificial intelligence to monitor and analyse online behaviour, often to personalise content, advertisements, and user experiences.

Algorithm – A set of rules or calculations that social media and search engines use to determine what content users see.

Algorithmic Distraction – The process by which AI-driven algorithms curate and suggest content to maximise user engagement, often overriding conscious decision-making.

Algorithm-driven feeds – Content recommendations generated by artificial intelligence based on user behaviour, preferences, and engagement patterns.

Allotment Gardening – A type of small-scale gardening where individuals cultivate plants in a designated communal space, often associated with improved mental well-being and stress reduction.

Artificial Intelligence (AI) – The simulation of human intelligence in machines that can perform tasks such as learning, problem-solving, and decision-making.

Asynchronous Communication – A method of communication where responses are not required in real-time, allowing flexibility in work and collaboration.

Attention Economy – The concept that online platforms compete for users' attention as a valuable resource.

Attention Hijacking – The deliberate design of digital experiences to capture and hold a user's focus, often at the expense of their productivity or well-being.

Attention Residue: The lingering cognitive effect after switching tasks, making it challenging to fully concentrate on the new task.

Attention Restoration Theory (ART) – A psychological theory suggesting that exposure to nature helps replenish cognitive resources, improving focus, creativity, and mental clarity.

Autoplay features – A setting in digital platforms that automatically plays the next video, episode, or piece of content to keep users engaged.

Automation – The use of technology to perform tasks without human intervention, often replacing manual labour.

Autoplay – A media feature that automatically starts the next video, reducing user decision-making and increasing consumption.

B

Black Box AI - An AI system whose decision-making process is not easily interpretable or understood by humans, raising concerns about transparency and accountability.

Bias (Algorithmic Bias) – A systematic error in an algorithm that leads to unfair outcomes, often reflecting societal prejudices.

Blockchain - A decentralised and distributed digital ledger technology that records transactions across multiple computers securely and transparently.

Biophilic Design – An architectural and interior design approach that incorporates natural elements, such as plants, natural lighting, and organic materials, to improve well-being and productivity.

Blue Light – A high-energy, short-wavelength light emitted by digital screens, which can interfere with melatonin production and sleep quality.

Burnout – A state of emotional, physical, and mental exhaustion caused by prolonged stress, often associated with overwork and lack of boundaries.

C

Cancel Culture – A social phenomenon where individuals or groups face widespread public backlash and ostracisation, often due to perceived offensive or problematic behaviour, typically amplified by social media.

Circadian Rhythms – The body's internal clock that regulates sleep-wake cycles, influenced by natural light and disrupted by exposure to blue light from screens.

Cognitive Load – The amount of mental effort required to process information, which can be overwhelmed by excessive digital stimuli.

Cognitive Overload – The mental exhaustion that occurs when too much information is processed at once, reducing focus, decision-making ability, and emotional regulation.

Conscious Consumption – The act of being mindful and intentional about the type and amount of digital content consumed.

Cortisol – A hormone produced by the adrenal glands that plays a key role in the body's stress response. Elevated cortisol levels are associated with anxiety, poor sleep, and other health issues.

Credibility Check – The process of verifying the accuracy and trustworthiness of digital content before accepting or sharing it.

Curated Content – Online material that has been selectively edited or presented in a way that highlights certain aspects while omitting others, often creating an unrealistic portrayal of reality.

Cryptography – The practice of securing communication and data through encoding techniques, used in wartime and modern cybersecurity.

Cyberbullying – The use of digital communication tools to harass, intimidate, or harm individuals, often through repeated hostile behaviour.

D

Data Privacy – The right of individuals to control how their personal data is collected, stored, and used.

Decentralisation - The process of distributing control and decision-making away from a central authority, often associated with blockchain and Web3 technologies.

Deepfake – A form of synthetic media that uses artificial intelligence to manipulate or generate realistic images, videos, or audio, often making it difficult to distinguish between real and altered content.

Deep-Focus Sessions – Dedicated periods of uninterrupted concentration on a task, free from digital distractions.

Digital Amnesia: The tendency to forget information readily available online, as individuals rely on digital devices for memory storage.

Digital Boundaries – Limits set around technology use, including screen time management, online communication habits, and personal data protection, to promote a balanced digital lifestyle.

Digital Consumption – The act of engaging with digital content, such as social media, news, and videos, which can be passive or intentional.

Digital Contracts – Agreements (formal or informal) that outline expectations for technology use, such as communication boundaries in professional or personal settings.

Digital Detox – A period of refraining from using digital devices to reset habits and reduce screen dependency.

Digital distractions – Any form of interruption or loss of focus caused by digital devices, notifications, or online content.

Digital Ecosystem – The interconnected network of digital platforms, applications, and users that shape the online environment.

Digital Energy – The impact that digital interactions have on an individual's emotional and mental state, akin to the way human interactions influence mood and well-being.

Digital Engagement – The ways in which individuals interact with digital platforms, including social media, news, and online communities.

Digital Ethics – The principles and guidelines governing responsible and respectful behaviour in digital environments, including privacy, boundaries, and online interactions.

Digital Eye Strain (Computer Vision Syndrome) – A condition caused by prolonged screen use, leading to symptoms such as headaches, dry eyes, blurred vision, and neck pain due to poor posture.

Digital Fatigue – Mental exhaustion resulting from prolonged exposure to digital stimuli, leading to reduced concentration, motivation, and emotional resilience.

Digital Identity - A unique, verifiable representation of a person, organisation, or entity in the digital world, often used for authentication and access control.

Digital Literacy – The ability to navigate digital platforms responsibly, understand how algorithms influence online experiences, and critically assess digital content for accuracy and ethical use.

Digital Mindfulness – The practice of using technology intentionally and with awareness to reduce stress, avoid distractions, and maintain a healthy relationship with digital platforms.

Digital Minimalism – A philosophy of reducing unnecessary technology use to focus on meaningful engagement.

Digital Overload – A state of mental exhaustion caused by excessive digital consumption, often leading to decreased productivity and well-being.

Digital Overstimulation – A state of cognitive and sensory overload caused by excessive exposure to digital devices, leading to stress, fatigue, and reduced attention span.

Digital Overwhelm – The feeling of stress, fatigue, or anxiety caused by excessive exposure to digital information and technology.

Digital Pile-On – A situation in which multiple individuals engage in online harassment or criticism of a single person, often escalating quickly and leading to significant emotional distress.

Digital Priorities – The aspects of digital engagement that a person values most, such as mental health, productivity, or social connections.

Digital Reflection – The practice of analysing one's digital habits and interactions to improve the quality of technology use.

Digital Surveillance – The monitoring of online activities, often conducted by governments or corporations to track behaviour.

Digital Trigger – Online content or interactions that provoke strong emotional reactions, such as anger, fear, or sadness, often leading to impulsive responses.

Digital Toolkit – A personalised set of strategies, apps, and guidelines designed to improve well-being and ethical engagement with technology.

Digital Validation – The act of seeking approval, recognition, or self-worth through online engagement, such as likes, comments, and shares.

Digital Well-Being – A balanced and intentional approach to using technology that supports mental, emotional, and physical health.

Doomscrolling – The act of continuously consuming negative news and online content, often leading to heightened anxiety and distress.

Dopamine – A neurotransmitter linked to pleasure, motivation, and addiction, often triggered by digital interactions.

Dopamine Desensitisation – A process where excessive dopamine stimulation reduces sensitivity to rewards, making previously enjoyable activities feel less satisfying.

Dopamine Loops: Behavioural cycles reinforced by dopamine release, such as the habitual checking of devices due to notifications.

Doxxing – The act of publicly revealing personal or private information about an individual, usually without their consent, to intimidate or harm them.

E

Echo Chamber – An online environment where individuals are exposed only to information that reinforces their existing beliefs.

Emotional Contagion – The phenomenon where emotions spread from one person to another, particularly in digital interactions.

Emotional Perception – The way individuals interpret and react to emotions in digital content, which can be influenced by the way information is presented online.

Emotional Regulation – The ability to manage and control one's emotional responses, particularly in high-stress situations such as online interactions.

Endless Scroll – A design feature that continuously loads new content as a user scrolls, eliminating natural stopping points and encouraging prolonged engagement.

Engagement Loop – A behavioural cycle where digital platforms use rewards (likes, comments, notifications) to reinforce continued use and interaction.

Environmental Psychology – A field of psychology that studies the relationship between humans and their surroundings, including how natural and built environments affect emotions and behaviour.

Ergonomics – The science of designing a workspace to minimise strain on the body, including proper posture, screen positioning, and chair height.

Ethical Accountability in Technology – The responsibility of individuals and organisations to engage with digital tools and platforms in a way that promotes positive online behaviour and ethical considerations.

Ethical Engagement – The practice of interacting online with awareness, responsibility, and respect for others, ensuring that digital spaces remain constructive and safe.

Ethical Framework – A structured approach to evaluating moral and ethical considerations in decision-making, especially in technology.

Ethical Innovation – The practice of developing technology with a focus on social responsibility, transparency, and user well-being.

Ethical Tech Use – The practice of engaging with digital platforms in a responsible and considerate manner, ensuring privacy, fairness, and well-being.

Ethical Technology Engagement – The practice of using technology in a way that is responsible, mindful, and aligned with human well-being.

Explainable AI (XAI) - A field of AI research focused on creating models that provide clear and understandable explanations for their decisions.

F

Fairness in AI - The principle that AI systems should produce unbiased and equitable outcomes, avoiding discrimination based on factors such as race, gender, or socioeconomic status.

Forest Bathing (Shinrin-yoku) – A Japanese practice that involves immersing oneself in a forest environment to engage the senses and promote relaxation and well-being.

G

Green Space – Any natural area, such as parks, forests, or gardens, that provides environmental and psychological benefits to people.

H

Habit Loop – A psychological pattern that consists of a cue, craving, response, and reward, reinforcing behaviours through repetition.

Highlight Reel Effect – The tendency for social media users to post only their best moments, leading to unrealistic comparisons and distorted perceptions of reality.

Hyper-Connectivity – A state of constant digital engagement through multiple devices, often leading to information overload and reduced mindfulness.

Hyper-fixation – An intense and prolonged focus on a specific digital task, often to the exclusion of other responsibilities, which can be both beneficial and detrimental depending on context and duration.

I

Inbox Detox – A strategy for managing email overload by setting specific times to check emails, unsubscribing from unnecessary newsletters, and using filters to prioritise important messages.

Industrial Revolution – A period of major industrialisation from the 18th to 19th centuries that drastically changed society through mechanisation.

Infinite Scrolling – A digital design feature that removes stopping cues, making users engage with content for longer periods.

Information Revolution – A shift in the accessibility and dissemination of information, such as the printing press and the internet.

Integrated Circuit – A foundational technology in computing, allowing multiple electronic components to be placed on a single chip.

Intentional Digital Engagement – A conscious approach to using digital platforms by setting boundaries, curating content, and engaging meaningfully rather than impulsively.

Intermittent Reinforcement – A conditioning strategy where rewards are given unpredictably, increasing addictive behaviours (e.g., slot machines, social media notifications).

M

Mass Production – The large-scale manufacturing of goods using assembly lines and automated systems.

Mental Well-being – A state of psychological health that includes emotional stability, cognitive clarity, and overall life satisfaction.

Metaverse - A collective virtual shared space, created by the convergence of virtually enhanced physical and digital reality, where users interact in real time.

Mindful Consumption – The practice of deliberately choosing digital content that is enriching, rather than engaging in passive or negative scrolling.

Mindful Digital Consumption – The practice of intentionally engaging with digital content in a way that aligns with personal goals, values, and well-being.

Mindful Digital Diet – A structured approach to balancing digital consumption across educational, creative, and entertainment content to enhance well-being.

Mindful Digital Engagement – The practice of consciously choosing how to interact with technology rather than engaging in habitual or reactive use.

Mindful Engagement – The practice of consciously interacting with digital content in a way that promotes emotional well-being and thoughtful discourse, rather than reactive or compulsive behaviour.

Mindful Technology Use – Conscious and intentional use of digital devices to minimise distractions and maximise benefits.

Mindfulness – A mental practice of focusing one's awareness on the present moment, often through breathing techniques, meditation, or conscious observation, to reduce stress and enhance emotional resilience.

Misinformation – False or misleading information spread, intentionally or unintentionally, through digital and social media platforms.

N

Neuroplasticity – The brain's ability to reorganise and adapt by forming new neural connections in response to experiences and learning.

Neuroscience – The scientific study of the nervous system, including the brain and its influence on behaviour and cognitive functions.

O

Online Privacy – The practice of protecting personal information and data shared on digital platforms, including managing privacy settings, securing sensitive information, and being mindful of digital footprints.

Online Triggers – Digital content, interactions, or notifications that provoke strong emotional responses, often leading to stress, frustration, or compulsive behaviours.

Outrage Culture – A digital phenomenon where individuals or groups react strongly and emotionally to online content, often leading to widespread debates or conflicts.

Oxytocin – Often called the "bonding hormone" or "love hormone," it is released through physical touch and social interactions, playing a key role in trust, attachment, and emotional warmth.

P

Parasympathetic Nervous System – Part of the autonomic nervous system responsible for relaxation and recovery, often activated through nature exposure and deep breathing.

Participatory Culture – A concept (introduced by *Henry Jenkins)* describing a digital environment where users actively create, share, and shape content rather than just consuming it.

Passive consumption – The unconscious or automatic use of digital content without intentional decision-making.

Passive Consumption Loop – A behavioural pattern where individuals continuously consume digital content (e.g., binge-watching or infinite scrolling) without conscious decision-making.

Persuasive Technology – Digital tools and design strategies (e.g., notifications, infinite scrolling) intended to influence users' behaviour *(Fogg, 2003).*

Polarisation – The process by which individuals or groups develop extreme opinions due to exposure to highly charged or one-sided digital content.

Pomodoro Technique: A time management method involving 25-minute focused work sessions followed by short breaks.

Predictive Analytics – The use of data and algorithms to anticipate future behaviours or outcomes.

Privacy – The right of individuals to control access to their personal information and maintain confidentiality in digital interactions.

Privacy and Consent – The principles of maintaining personal boundaries in digital spaces, ensuring transparency in data sharing, and obtaining explicit permission before sharing personal information
Privacy Audit – A self-assessment of how one's personal data is being collected, used, and shared by digital services.

Propaganda – Information, often biased or misleading, used to influence public opinion or behaviour.

Push Notifications – Alerts sent by apps to notify users of updates, messages, or activity, often creating a sense of urgency to re-engage.

R

Reactive Loop – A cycle where notifications, alerts, and social media interactions trigger automatic emotional responses, often leading to stress or compulsive engagement.

Recommendation Engine - An AI-based system that analyses user data to suggest personalised content, such as products, movies, or news articles.

Reflection Exercise – A guided mental practice encouraging individuals to analyse their experiences and behaviours to promote self-awareness and growth.

Right to Disconnect – A policy or legal framework that protects employees from being required to engage in workrelated communication outside of designated working hours.

S

Salivary Biomarkers – Biological indicators found in saliva, such as cortisol, used to measure stress and other physiological responses.

Screen Exposure – The amount of time spent looking at digital screens, which can impact sleep, focus, and overall health.

Screen Time – The amount of time spent using digital devices such as smartphones, tablets, or computers.

Screen Time Audit – A process of reviewing and analysing time spent on various digital activities to assess their impact on productivity and well-being.

Self-Regulation – The ability to control one's online behaviour, including limiting screen time, managing emotional responses, and engaging with content mindfully.

Single-tasking – The practice of focusing on one task at a time rather than multitasking, leading to improved productivity and mindfulness.

Social Connections (in a digital context) – The relationships and interactions a person maintains through digital communication tools.

Social Disconnection – A phenomenon where excessive digital engagement leads to weakened real-world relationships and diminished face-to-face communication skills.

Social Media Platform – An online service or website that enables users to create, share, and interact with content and other users.

Social Validation – The psychological need to be accepted and affirmed by others, often exploited by social media platforms through features such as likes, comments, and shares.

Social Validation Loop – A cycle where online engagement (likes, comments) reinforces posting behaviours through dopamine-driven feedback.

Stress Recovery Theory – A psychological theory suggesting that exposure to natural environments helps reduce stress and promote emotional well-being.

Surveillance Capitalism – A term describing how companies collect and monetise user data for targeted advertising and influence.

Sustainability in Technology – The principle of developing and using digital tools in ways that minimise environmental harm and promote long-term viability.

T

Tech Boundaries – Limits set around technology use to maintain a healthy relationship with digital tools.

Tech Ethics – The study of moral issues related to technology development, implementation, and impact on society.

Tech-Free Zones – Physical spaces designated as free from digital devices, such as bedrooms or dining areas, to encourage real-world interactions.

Tech Accountability – The responsibility of both individuals and companies to ensure that digital tools do not cause harm and are used ethically.

Technology Addiction – A compulsive dependency on digital devices or online platforms, leading to negative consequences in daily life.

Technology Engagement – The ways in which a person interacts with digital tools, including social media, apps, and online platforms.

Technology-Induced Hyper-focus – A state of deep concentration triggered by engaging digital tasks, which can lead to exhaustion and neglect of real-world responsibilities.

Tech Transparency – The extent to which technology companies disclose their data practices, algorithms, and business models to the public.

The 60-30 Rule – A digital mindfulness practice where individuals focus deeply on tasks for 60 minutes, followed by a 30-minute break to reduce mental fatigue.

Time Inconsistency – A cognitive bias where people prioritise immediate rewards over long-term benefits.

U

Ultradian Rhythms – Biological cycles that govern periods of focus and rest, typically lasting around 90 minutes, influencing cognitive performance and energy levels.

Urban Nature – Natural elements found in city environments, such as trees, green roofs, and parks, which contribute to mental and physical health benefits.

User Autonomy – The ability of individuals to make informed decisions about how they engage with technology and digital platforms.

User Manipulation – The use of persuasive design and dark patterns to influence user behaviour in ways that may be unethical or harmful.

V

Virtual Economy - A digital marketplace where users can buy, sell, and trade virtual goods, services, and currency, often found in online games and the metaverse.

W

Work-Life Balance – The equilibrium between professional responsibilities and personal life, ensuring well-being and preventing burnout.

RESOURCES

A

Ainslie, G. (1975). Specious reward: A behavioural theory of impulsiveness and impulse control. Psychological Bulletin, 82(4), 463–496. https://doi.org/10.1037/h0076860

Alter, A. (2017). Irresistible: The rise of addictive technology and the business of keeping us hooked. Penguin Press.

American Optometric Association. (n.d.). Computer vision syndrome. Retrieved from www.aoa.org

American Psychological Association. (2006). Multitasking: Switching costs. Retrieved from https://www.apa.org/research/action/multitask

American Psychological Association. (2020). Stress in America™ 2020: A National Mental Health Crisis. https://www.apa.org/news/press/releases/stress/2020/sia-mental-health-crisis.pdf

Asurion. (2019). Americans check their phones 96 times a day. Retrieved from https://www.asurion.com

Atchley, R. A., Strayer, D. L., & Atchley, P. (2012). Creativity in the wild: Improving creative reasoning through immersion in natural settings. PLOS ONE, 7(12), e51474. https://doi.org/10.1371/journal.pone.0051474

Australian Psychological Society. (2020). Cyberbullying among adults. Retrieved from https://psychology.org.au/formembers/publications/inpsych/2020/april-may-issue-2/cyberbullying-among-adults

B Bailey, C. (2018). Hyperfocus: How to work less and achieve more. Viking Press.

Bak-Coleman, J. B., et al. (2021). Stewardship of global collective behaviour. Proceedings of the National Academy of Sciences, 118(27), e2025764118. https://doi.org/10.1073/pnas.2025764118

Brady, W. J., Wills, J. A., Jost, J. T., Tucker, J. A., & Van Bavel, J. J. (2017). Emotion shapes the diffusion of moralized content in social networks. Proceedings of the National Academy of Sciences, 114(28), 7313-7318. https://doi.org/10.1073/pnas.1618923114

Bratman, G. N., Hamilton, J. P., & Daily, G. C. (2015). The impacts of nature experience on human cognitive function and mental health. Annals of the New York Academy of Sciences, 1362(1), 1–15.

Benton, D. (2020). Digital Waste and Sustainability: The Environmental Costs of Our Online Lives. Oxford University Press.

Berman, M. G., Jonides, J., & Kaplan, S. (2008). The cognitive benefits of interacting with nature. Psychological Science, 19(12), 1207–1212.

Berridge, K. C., & Robinson, T. E. (1998). The role of dopamine in reward: The incentive sensitization theory of addiction. *Brain Research Reviews, 28*(3), 309–369. https://doi.org/10.1016/S0165-0173(98)00019-8

Bohn, R. E., & Short, J. E. (2009). How much information? 2009 report on American consumers. University of California, San Diego. Retrieved from https://hmi.ucsd.edu/pdf/HMI_2009_ConsumerReport_Dec9_2009.pdf

Brey, P. (2017). Ethics of emerging technologies: An introduction. Routledge.

Bridle, J. (2018). New Dark Age: Technology and the End of the Future. Verso Books.

Brown, L., & Miller, T. (2020). Digital Overload and Mental Well-being: The Psychological Cost of Online Spaces. Oxford University Press.

Brynjolfsson, E., & McAfee, A. (2014). The second machine age: Work, progress, and prosperity in a time of brilliant technologies. W. W. Norton & Company.

C

Cacioppo, J. T., & Cacioppo, S. (2014). Social relationships and health: The toxic effects of perceived social isolation. Social and Personality Psychology Compass, 8(2), 58–72. https://doi.org/10.1111/spc3.12087 Carr, N. (2010). The shallows: What the Internet is doing to our brains. W. W. Norton & Company.

Castronova, E. (2005). Synthetic worlds: The business and culture of online games. University of Chicago Press.

Chang, A. M., Aeschbach, D., Duffy, J. F., & Czeisler, C. A. (2015). Evening use of light-emitting eReaders negatively affects sleep, circadian timing, and next-morning alertness. Proceedings of the National Academy of Sciences, 112(4), 1232-1237. https://doi.org/10.1073/pnas.1418490112

Chesney, R., & Citron, D. (2019). Deepfakes and the new disinformation war: The coming age of post-truth geopolitics. Foreign Affairs, 98(1), 147–155.

Chevalier, G., Sinatra, S. T., Oschman, J. L., & Delany, R. M. (2012). Earthing: Health implications of reconnecting the human body to the Earth's surface electrons. Journal of Environmental and Public Health, 2012, 291541. https://doi.org/10.1155/2012/291541

Christakis, D. A. (2019). The effects of digital media on the brain: Science and policy. Pediatrics, 143(2), 1-10.

Chou, H. T. G., & Edge, N. (2012). "They are happier and having better lives than I am": The impact of using Facebook on perceptions of others' lives. Cyberpsychology, Behavior, and Social Networking, 15(2), 117-121. https://doi.org/10.1089/cyber.2011.0324

Cirillo, F. (2006). The Pomodoro Technique. FC Media.

D

Dastin, J. (2018, October 10). Amazon scraps secret AI recruiting tool that showed bias against women. Reuters. https://www.reuters.com

Davidson, R. J., & Begley, S. (2012). The emotional life of your brain: How its unique patterns affect the way you think, feel, and live—and how you can change them. Hudson Street Press.

Dignum, V. (2019). Responsible artificial intelligence: Developing and using AI in a responsible way. Springer.

Dunbar, R. I. M. (2010). The social role of touch in humans and primates: Behavioural function and neurobiological mechanisms. Neuroscience & Biobehavioral Reviews, 34(2), 260–268.

E

Eisenstein, E. L. (1980). The printing press as an agent of change. Cambridge University Press.

Eyal, N. (2014). Hooked: How to build habit-forming products. Portfolio/Penguin.

Eyal, N. (2019). Indistractable: How to Control Your Attention and Choose Your Life. BenBella Books.

F

Fair Work Amendment (Right to Disconnect) Act 2023 (Cth) (Austl.). https://www.legislation.gov.au

Field, T. (2010). Touch for socioemotional and physical well-being: A review. Developmental Review, 30(4), 367– 383.

Fardouly, J., Diedrichs, P. C., Vartanian, L. R., & Halliwell, E. (2015). Social comparisons on social media: The impact of Facebook on young women's body image concerns and mood. Body Image, 13, 38-45. https://doi.org/10.1016/j.bodyim.2014.12.002

Floridi, L. (2013). The ethics of information. Oxford University Press.

Fogg, B. J. (2003). Persuasive technology: Using computers to change what we think and do. Morgan Kaufmann.

Foxman, M. (2022). Virtual economies, real injustices: Economic disparities in the metaverse. Journal of Digital Ethics, 10(2), 220-235.

Freeman, C., & Louca, F. (2001). As time goes by: From the industrial revolutions to the information revolution. Oxford University Press.

French Labour Code. (2017). Loi n° 2016-1088 du 8 août 2016 relative au travail, à la modernisation du dialogue social et à la sécurisation des parcours professionnels [Law No. 2016-1088 of August 8, 2016, on labor, the modernization of social dialogue, and the securing of career paths]. https://www.legifrance.gouv.fr

G

Garcia, P. (2022). Interpreting Online Communication: The Missing Elements of Digital Conversations. Harvard University Press.

Gazzaley, A., & Rosen, L. D. (2016). The distracted mind: Ancient brains in a high-tech world. MIT Press.

Ghaly, M., & Teplitz, D. (2004). The biologic effects of grounding the human body during sleep as measured by cortisol levels and subjective reporting of sleep, pain, and stress. The Journal of Alternative and Complementary Medicine, 10(5), 767–776. https://doi.org/10.1089/acm.2004.10.767

Goleman, D., & Davidson, R. J. (2017). Altered Traits: Science Reveals How Meditation Changes Your Mind, Brain, and Body. Avery Publishing.

Gummerus, J., Liljander, V., Weman, E., & Pihlström, M. (2012). Customer engagement in a Facebook brand community. Management Research Review, 35(9), 857-877.

H

Haidt, J., & Allen, N. (2020). Screen time and the mental health of children and adolescents: Evidence, insights, and implications. Nature Human Behaviour, 4(2), 160–168. https://doi.org/10.1038/s41562-019-0784-7

Haidt, J., & Twenge, J. (2022). The Anxious Generation: How Social Media Creates Anxiety and What We Can Do About It. Oxford University Press.

Harari, Y. N. (2015). Sapiens: A brief history of humankind. Harper.

Harris, T. (2019). The Social Dilemma [Documentary]. Exposure Labs.

Harris, T. (2019). The attention economy and the human cost of social media. Journal of Digital Ethics.

Harvard Health Publishing. (2024). The benefits of nature for stress relief. Harvard Medical School. https://www.health.harvard.edu/mind-and-mood/the-benefits-of-nature-for-stress-relief

Harvard Health. (2024). The impact of nature on stress reduction and mental resilience. Harvard Medical School. Retrieved from https://www.health.harvard.edu

Harvard Medical School. (2012). Blue light has a dark side. Harvard Health Publishing. Retrieved from www.health.harvard.edu

Hatfield, E., Cacioppo, J. T., & Rapson, R. L. (1994). Emotional contagion. Cambridge University Press.

Hjalmarsson, A., Håkansson, J., & Conner, J. (2018). Blockchain-based e-voting systems: Opportunities and challenges. ACM Digital Library.

Holt-Lunstad, J., Smith, T. B., & Layton, J. B. (2010). Social relationships and mortality risk: A meta-analytic review. PLoS Medicine, 7(7), e1000316. https://doi.org/10.1371/journal.pmed.1000316

Howard, P. N., Woolley, S. C., & Calo, R. (2021). Lies, damn lies, and algorithms: The impact of AI-driven misinformation. Oxford University Press.

Hunter, M. R., Gillespie, B. W., & Chen, S. Y. P. (2019). Urban nature experiences reduce stress in the context of daily life based on salivary biomarkers. Frontiers in Psychology, 10, 722.

J

Jenkins, H. (2006). Convergence Culture: Where Old and New Media Collide. NYU Press.

Johnson, S. (2019). The innovators: How a group of hackers, geniuses, and geeks created the digital revolution. Penguin Books.

Johnson, M., & Lee, S. (2020). The impact of digital clutter on cognitive load and mental fatigue. Journal of Digital Well-being, 15(2), 134-148.

K

Kaspersky Lab. (2015). The rise of digital amnesia: Why we need to protect what we no longer remember. Retrieved from https://www.kaspersky.com/blog/digital-amnesia

Kobie, N. (2021, August 10). China's AI judges are deciding more and more court cases. Wired. https://www.wired.com

Kramer, A. D. I., Guillory, J. E., & Hancock, J. T. (2014). Experimental evidence of massive-scale emotional contagion through social networks. Proceedings of the National Academy of Sciences, 111(24), 8788-8790. https://doi.org/10.1073/pnas.1320040111

Kranzberg, M. (1986). Technology and history: "Kranzberg's laws". Technology and Culture, 27(3), 544-560.

L

Laibson, D. (1997). Golden eggs and hyperbolic discounting. The Quarterly Journal of Economics, 112(2), 443–477. https://doi.org/10.1162/003355397555253

Lee, C. (2019). Mindful Technology: A Guide to Ethical and Intentional Digital Engagement. MIT Press.

Lembke, A. (2021). Dopamine nation: Finding balance in the age of indulgence. Dutton.

Leroy, S. (2009). "Why is it so hard to do my work? The challenge of attention residue when switching between work tasks." Organisational Behaviour and Human Decision Processes, 109(2), 168–181. https://doi.org/10.1016/j.obhdp.2009.04.002

Lewandowsky, S., Ecker, U. K. H., & Cook, J. (2017). Beyond misinformation: Understanding and coping with the "post-truth" era. Journal of Applied Research in Memory and Cognition, 6(4), 353-369. https://doi.org/10.1016/j.jarmac.2017.07.008

Lipton, Z. C. (2018). The mythos of model interpretability. ACM Queue, 16(3), 31–57.

Lyubomirsky, S., & Layous, K. (2013). How do simple positive activities increase well-being? Current Directions in Psychological Science, 22(1), 57-62. https://doi.org/10.1177/0963721412469809

M

Mark, G., Gonzalez, V. M., & Harris, J. (2005). No task left behind? Examining the nature of fragmented work. Proceedings of the SIGCHI Conference on Human Factors in Computing Systems, 321-330. https://doi.org/10.1145/1054972.1055017

Mark, G., Iqbal, S. T., Czerwinski, M., Johns, P., Sano, A., & Lutchyn, Y. (2016). Focused, aroused, but so distractible: A temporal perspective on multitasking and communications. Proceedings of the 2016 CHI Conference on Human Factors in Computing Systems, 903–916. https://doi.org/10.1145/2858036.2858202

Mehrabi, N., Morstatter, F., Saxena, N., Lerman, K., & Galstyan, A. (2021). A survey on bias and fairness in machine learning. ACM Computing Surveys, 54(6), 1–35.

Meyer, D. E., & Kieras, D. E. (1997). A computational theory of executive cognitive processes and multiple-task performance: Part 1, Basic mechanisms. Psychological Review, 104(1), 3-65. https://doi.org/10.1037/0033295X.104.1.3

Misa, T. J. (2011). Digital state: The story of Minnesota's computing industry. University of Minnesota Press.

Mittelstadt, B. D., Allo, P., Taddeo, M., Wachter, S., & Floridi, L. (2016). The ethics of algorithms: Mapping the debate. Big Data & Society, 3(2), 1-21.

N

Natasha Schüll's Addiction by Design (2012)

Newport, C. (2016). Deep work: Rules for focused success in a distracted world. Grand Central Publishing.

Newport, C. (2019). Digital Minimalism: Choosing a Focused Life in a Noisy World. Penguin Random House.

Noble, S. U. (2018). Algorithms of Oppression: How Search Engines Reinforce Racism. NYU Press.

O

O'Neil, C. (2016). Weapons of Math Destruction: How Big Data Increases Inequality and Threatens Democracy. Crown Publishing Group.

P

Pariser, E. (2011). The Filter Bubble: What the Internet Is Hiding from You. Penguin Press.

Park, B. J., Tsunetsugu, Y., Kasetani, T., Kagawa, T., & Miyazaki, Y. (2010). Physiological effects of Shinrin-yoku (forest bathing) and the implications for health promotion. Environmental Health and Preventive Medicine, 15(1), 27– 37.

Pasquale, F. (2015). The black box society: The secret algorithms that control money and information. Harvard University Press.

R

Radtke, T., Apel, T., Schenkel, K., Keller, J., & von Lindern, E. (2022). Digital detox: An effective solution in the smartphone era? Mobile Media & Communication, 10(2), 190–211. https://doi.org/10.1177/20501579211060130

Risko, E. F., & Gilbert, S. J. (2016). Cognitive offloading. Trends in Cognitive Sciences, 20(9), 676–688. https://doi.org/10.1016/j.tics.2016.07.002

Rosen, L. D., Mark, A. F., & Cheever, N. A. (2013). iDisorder: Understanding our obsession with technology and overcoming its hold on us. Palgrave Macmillan.

Russell, S. (2021). Human compatible: Artificial intelligence and the problem of control. Penguin Books.

S

Shanahan, D. F., Bush, R., Gaston, K. J., Lin, B. B., Dean, J., Barber, E., & Fuller, R. A. (2016). Health benefits from nature experiences depend on dose. Scientific Reports, 6, 28551.

Small, G., et al. (2020). This study explores how frequent use of digital technology may significantly impact adult attention spans.

Sparrow, B., Liu, J., & Wegner, D. M. (2011). Google effects on memory: Cognitive consequences of having information at our fingertips. Science, 333(6043), 776–778. https://doi.org/10.1126/science.1207745 Standage, T. (2013). Writing on the wall: Social media - The first 2,000 years. Bloomsbury.

Stanford Center on Longevity. (2024). The future of long life: How technology shapes aging and well-being. https://www.longevity.stanford.edu/future-of-long-life

Skinner, B. F. (1953). Science and Human Behavior. Macmillan.

Smith, J., & Doe, A. (2023). The impact of digital consumption on mental health. BMC Psychology, 11(4), 112-125. https://doi.org/10.xxxx

Smith, J., & Jones, R. (2022). The digital habit: How technology shapes our daily lives. Oxford University Press.

Smith, R., & Johnson, K. (2021). The Attention Economy: How Digital Platforms Shape Our Thinking and Feeling. Routledge.

Solove, D. J. (2020). The Digital Person: Technology and Privacy in the Information Age. NYU Press.

Stone, L. (2008, February 12). Beyond simple multitasking: Continuous partial attention. Linda Stone. https://lindastone.net/2008/02/12/beyond-simple-multitasking-continuous-partial-attention/

Suler, J. (2004). The online disinhibition effect. CyberPsychology & Behavior, 7(3), 321–326. https://doi.org/10.1089/1094931041291295

T

Turkle, S. (2015). Reclaiming conversation: The power of talk in a digital age. Penguin Books.

Turner, J. (2021). Emotional Contagion in the Digital Age: The Spread of Negativity and Positivity Online. Cambridge University Press.

Turner, K., Hancock, J. T., & Landay, J. A. (2021). Doomscrolling and anxiety: Examining the link between excessive online news consumption and mental health. Journal of Behavioral Addictions, 10(3), 522-535. https://doi.org/10.1556/2006.2021.00036

Twenge, J. M. (2017). iGen: Why today's super-connected kids are growing up less rebellious, more tolerant, less happy—and completely unprepared for adulthood. Atria Books.

Twenge, J. M., & Campbell, W. K. (2018). Associations between screen time and lower psychological well-being among children and adolescents: Evidence from a population-based study. Preventive Medicine Reports, 12, 271– 283.

U

Ulrich, R. S., Simons, R. F., Losito, B. D., Fiorito, E., Miles, M. A., & Zelson, M. (1991). Stress recovery during exposure to natural and urban environments. Journal of Environmental Psychology, 11(3), 201–230.

V

Volkow, N. D., Wang, G. J., Tomasi, D., & Baler, R. D. (2011). Obesity and addiction: Neurobiological overlaps. Obesity Reviews, 12(5), e106-e120. https://doi.org/10.1111/j.1467-789X.2010.00898.x

Vosoughi, S., Roy, D., & Aral, S. (2018). The spread of true and false news online. Science, 359(6380), 1146-1151. https://doi.org/10.1126/science.aap9559

W

Walker, M. (2017). Why we sleep: Unlocking the power of sleep and dreams. Scribner.

Ward, A. F., Duke, K., Gneezy, A., & Bos, M. W. (2017). Brain drain: The mere presence of one's own smartphone reduces available cognitive capacity. Journal of the Association for Consumer Research, 2(2), 140-154. https://doi.org/10.1086/691462

Whittaker, M., Crawford, K., Dobbe, R., Fried, G., Kaziunas, E., Mathur, V., ... & Schwartz, O. (2018). AI now report 2018. AI Now Institute, New York University.

Williams, E., Brown, J., & Smith, R. (2021). AI-driven content moderation in the metaverse: Opportunities and risks. Journal of Emerging Technologies, 15(4), 112–130.

Wood, E., Hooper, P., Foster, S., & Bull, F. (2018). The relationship between greenspace and the mental wellbeing of adults: A systematic review. PLOS ONE, 13(9), e0203000. https://doi.org/10.1371/journal.pone.0203000

World Health Organization. (2019). Public health implications of excessive screen time and digital addiction. WHO Report.

Williams, J. (2018). Stand Out of Our Light: Freedom and Resistance in the Attention Economy. Cambridge University Press.

Z

Zuboff, S. (2019). The age of surveillance capitalism: The fight for a human future at the new frontier of power. PublicAffairs.

Zyskind, G., Nathan, O., & Pentland, A. (2015). Decentralizing privacy: Using blockchain to protect personal data. Proceedings of the IEEE Security and Privacy Workshops, 180–184.

Made in the USA
Monee, IL
04 November 2026

67384790R00095